ADDITIONAL SKILL AND DRILL MANUAL

JAMES J. BALL
Indiana State University

INTRODUCTORY AND INTERMEDIATE ALGEBRA

THIRD EDITION

Margaret L. Lial
American River College

John Hornsby
University of New Orleans

Terry McGinnis

PEARSON

Addison
Wesley

Boston San Francisco New York
London Toronto Sydney Tokyo Singapore Madrid
Mexico City Munich Paris Cape Town Hong Kong Montreal

Reproduced by Pearson Addison-Wesley from electronic files supplied by the author.

Copyright © 2006 Pearson Education, Inc.
Publishing as Pearson Addison-Wesley, 75 Arlington Street, Boston, MA 02116.

ISBN   0-321-33169-9

2 3 4 5 6 OPM 08 07 06 05

CONTENTS

Chapter R

PREALGEBRA REVIEW

R.1 Fractions

Objective 1   Identify prime numbers.

Tell whether each number is prime, composite, or neither.

1. 29   **4.** 31   **7.** 3693   **10.** 301

2. 35   **5.** 23   **8.** 79   **11.** 2488

3. 1   **6.** 482   **9.** 127   **12.** 59

7.3693

Objective 2   Write numbers in prime factored form.

Write each number in prime factored form.

13. 68   **16.** 78   **19.** 189   **22.** 248

14. 45   **17.** 24   **20.** 256   **23.** 59

15. 98   **18.** 210   **21.** 168   **24.** 546

Objective 3   Write fractions in lowest terms.

Write each fraction in lowest terms.

25. $\dfrac{30}{56}$ **28.** $\dfrac{42}{12}$ **31.** $\dfrac{28}{140}$ **34.** $\dfrac{315}{225}$

26. $\dfrac{25}{35}$ **29.** $\dfrac{42}{150}$ **32.** $\dfrac{180}{216}$ **35.** $\dfrac{224}{504}$

27. $\dfrac{25}{70}$ **30.** $\dfrac{60}{38}$ **33.** $\dfrac{216}{416}$ **36.** $\dfrac{292}{132}$

Objective 4   **Multiply and divide fractions.**

Find each product, and write it in lowest terms.

37. $\dfrac{7}{3} \cdot \dfrac{9}{28}$

40. $\dfrac{3}{5} \cdot 20$

43. $\dfrac{5}{6} \cdot \dfrac{12}{20}$

46. $4\frac{3}{8} \cdot 2\frac{4}{7}$

47. $4\frac{1}{6} \cdot 3\frac{1}{5}$

38. $\dfrac{3}{8} \cdot \dfrac{12}{33}$

41. $\dfrac{25}{11} \cdot \dfrac{33}{10}$

44. $\dfrac{6}{7} \cdot \dfrac{21}{24}$

48. $5\frac{1}{4} \cdot 3\frac{1}{2}$

39. $\dfrac{4}{9} \cdot \dfrac{18}{8}$

42. $\dfrac{8}{18} \cdot \dfrac{9}{24}$

45. $4\frac{2}{3} \cdot 8\frac{1}{4}$

Find each quotient, and write it in lowest terms.

49. $\dfrac{5}{4} \div \dfrac{25}{28}$

52. $\dfrac{5}{7} \div \dfrac{40}{21}$

55. $\dfrac{4}{11} \div 32$

58. $4\frac{4}{9} \div 2\frac{2}{3}$

59. $7\frac{1}{2} \div 3\frac{1}{3}$

50. $\dfrac{3}{8} \div \dfrac{12}{56}$

53. $\dfrac{12}{13} \div 18$

56. $\dfrac{16}{20} \div \dfrac{24}{60}$

60. $4\frac{2}{3} \div 2\frac{1}{3}$

51. $\dfrac{3}{4} \div \dfrac{18}{24}$

54. $\dfrac{12}{11} \div \dfrac{48}{22}$

57. $3\frac{1}{3} \div 4\frac{2}{3}$

Objective 5   **Add and subtract fractions.**

Add. Write answers in lowest terms.

61. $\dfrac{3}{10} + \dfrac{5}{10}$

64. $\dfrac{5}{12} + \dfrac{7}{18}$

67. $\dfrac{13}{8} + \dfrac{7}{4}$

70. $2\frac{3}{4} + 7\frac{2}{3}$

71. $4\frac{3}{5} + 3\frac{1}{2}$

65. $\dfrac{3}{7} + \dfrac{5}{21}$

62. $\dfrac{2}{9} + \dfrac{4}{9}$

68. $\dfrac{11}{16} + \dfrac{7}{20}$

72. $14\frac{3}{4} + 26\frac{5}{8}$

63. $\dfrac{6}{15} + \dfrac{4}{15}$

66. $\dfrac{23}{45} + \dfrac{47}{75}$

69. $\dfrac{5}{12} + \dfrac{3}{4} + \dfrac{2}{9}$

Subtract. Write answers in lowest terms.

73. $\dfrac{7}{8} - \dfrac{5}{8}$

75. $\dfrac{11}{12} - \dfrac{3}{12}$

77. $\dfrac{5}{12} - \dfrac{1}{8}$

79. $\dfrac{8}{30} - \dfrac{2}{20}$

74. $\dfrac{12}{17} - \dfrac{7}{17}$

76. $\dfrac{8}{9} - \dfrac{3}{4}$

78. $\dfrac{7}{15} - \dfrac{2}{9}$

80. $\dfrac{17}{18} - \dfrac{1}{4}$

81. $\dfrac{21}{16} - \dfrac{6}{9}$ **82.** $10\frac{1}{9} - 4\frac{2}{3}$ **83.** $11\frac{1}{6} - 7\frac{2}{3}$ **84.** $128\frac{5}{16} - 57\frac{3}{8}$

R.1 Mixed Exercises

Write each number in prime factored form.

85. 150   **87.** 47   **89.** 420

86. 336   **88.** 1500   **90.** 310

Perform indicated operations. Write answers in lowest terms.

91. $\dfrac{4}{15} + \dfrac{6}{5}$ **94.** $\dfrac{2}{9} \div \dfrac{21}{36}$ **97.** $\dfrac{11}{6} \div \dfrac{77}{24}$ **100.** $12\frac{7}{8} - 7\frac{5}{6}$

101. $6\frac{2}{3} \cdot 8\frac{1}{4}$

92. $\dfrac{13}{15} - \dfrac{3}{5}$ **95.** $\dfrac{11}{13} + \dfrac{15}{39}$ **98.** $\dfrac{21}{45} - \dfrac{5}{18}$ **102.** $8\frac{4}{5} \div 4\frac{2}{5}$

93. $\dfrac{75}{20} \cdot \dfrac{24}{15}$ **96.** $\dfrac{7}{8} \cdot \dfrac{24}{28}$ **99.** $4\frac{1}{5} + 1\frac{11}{12}$

PREALGEBRA REVIEW

R.2 Decimals and Percents

Objective 1   Write decimals as fractions.

Write each decimal as a fraction. Do not write in lowest terms.

1. .3	**4.** .247	**7.** .007	**10.** 18.6
2. .42	**5.** .234	**8.** .092	**11.** 4.28
3. .26	**6.** .427	**9.** .0054	**12.** 8.653

Objective 2   Add and subtract decimals.

Add or subtract as indicated.

13. 28.4
   + 11.08

14. 212.43
   + 2.573

15. 428.24
   12.91
   + 39.65

16. 284.381
   − 28.292

17. 32.006
   − 2.492

18. 768.5
   − 13.402

19. $7.2 + 2.45 + 6.7$

20. $128.1 - 20.3$

21. $7.26 + 3.7 + 4.09$

22. $694.2 - 538.9$

23. $27.39 + 9.7 + 28.24$

24. $405.2 - 314.48 + 29.029$

Objective 3   Multiply and divide decimals.

Multiply or divide as indicated. Round to the nearest thousandth.

25. 3.2×12.5	**29.** $45.8 \times .09$	**33.** $498.72 \div 21.2$
26. 35×92.46	**30.** $48.93 \times .87$	**34.** $516.65 \div 12.8$
27. $23.5 \times .42$	**31.** $38.6 \div .05$	**35.** $248.92 \div 14.6$
28. $14.64 \times .16$	**32.** $76.24 \div .12$	**36.** $48.5 \div .002$

37. 24.8×10

38. 42.789×100

39. 467.04×1000

40. $45.7 \div 10$

41. $394.8 \div 100$

42. $2.687 \div 100$

43. 36.421×1000

44. $429.2 \div 1000$

45. $.384 \times 10,000$

46. $7.4102 \div 1000$

Objective 4   **Write fractions as decimals.**

Write each fraction as a decimal. Round to the nearest thousandth, and for repeating decimals, write the answer two ways: using the bar notation and rounding to the nearest thousandth.

47. $\dfrac{3}{7}$

48. $\dfrac{3}{10}$

49. $\dfrac{9}{16}$

50. $\dfrac{5}{8}$

51. $\dfrac{4}{11}$

52. $\dfrac{13}{6}$

53. $\dfrac{4}{9}$

54. $\dfrac{41}{40}$

55. $\dfrac{4}{17}$

56. $\dfrac{5}{54}$

57. $\dfrac{151}{200}$

58. $\dfrac{1}{3}$

Objective 5   **Convert percents to decimals and decimals to percents.**

Convert the following percents to decimals.

59. 41%

60. 29%

61. 13%

62. $.4\%$

63. $.28\%$

64. 1.4%

65. 27.5%

66. 100%

67. 362%

Convert the following decimals to percents.

68. $.24$

69. $.23$

70. $.045$

71. 9

72. 2.35

73. $.004$

74. 2.09

75. $.0409$

76. $.0084$

R.2 Mixed Exercises

Perform indicated operations.

77.   425.371
       + 219.29

78.   748.2
       − 564.28

79. $9.02 + 7.4 + 4.46$

80. $340.2 - 46.8$

81. $402.9 - 25.01$

82. 45.2×48.2

83. 241×2.18

84. $34.1 \times .07$

85. $8.08 \div .07$

86. $147.94 \div 2.02$

87. 5.2×1000

88. $35.04 \div 1000$

Convert as indicated.

89. .84 to a fraction

90. $\frac{3}{16}$ to a decimal

91. 49% to a decimal

92. .042 to a percent

93. $\frac{4}{11}$ to a decimal

94. $\frac{13}{6}$ to a percent

95. $\frac{7}{15}$ to a decimal

96. 2.45 to a percent

97. .27 to a fraction

98. $\frac{8}{3}$ to a percent

Chapter 1

THE REAL NUMBER SYSTEM

1.1 Exponents, Order of Operations, and Inequality

Objective 1 Use exponents.

Find the value of each exponential expression.

1. 3^3

2. 2^5

3. 3^4

4. 2^4

5. $\left(\dfrac{3}{4}\right)^2$

6. $\left(\dfrac{1}{2}\right)^6$

7. $\left(\dfrac{2}{5}\right)^3$

8. $\left(\dfrac{2}{3}\right)^4$

9. $(.4)^2$

10. $(.03)^3$

11. $(.9)^2$

12. $(2.4)^3$

Objective 2 Use the order of operations guidelines.

Find the value of each expression.

13. $3 \cdot 4 + 5$

14. $7 + 2 \cdot 4$

15. $5 + 16 \div 4$

16. $3 \cdot 15 - 10^2$

17. $9 \cdot 3 + 3 \cdot 6$

18. $20 \div 5 - 3 \cdot 1$

19. $\dfrac{1}{2} \cdot \dfrac{2}{3} + \dfrac{3}{4} \cdot \dfrac{1}{3}$

20. $\left(\dfrac{5}{6}\right)\left(\dfrac{3}{2}\right) - \left(\dfrac{1}{3}\right)^2$

21. $(1.2)(2.3) - (.4)(.8)$

22. $\dfrac{4 \cdot 10 + 9 \cdot 2}{2(4-2)}$

23. $\dfrac{3 \cdot 15 + 10^2}{12^2 - 8^2}$

24. $\dfrac{10(5-3) - 9(6-2)}{2(4-1) - 2^2}$

Objective 3   Use more than one grouping symbol.

Find the value of each expression.

25. $6\left[5+3(4)\right]$

26. $10+4\left[2(6)-3\right]$

27. $8+4\left[2(4)+3\right]$

28. $2\left[6(5)-8\right]$

29. $4+3\left[7-3(3)\right]$

30. $9\left[2(5)-4\right]+7$

31. $4\left[3+2(9-2)\right]$

32. $8\left[14-3(9-4)\right]$

33. $19-3\left[8(5-2)+6\right]$

34. $4\left[5+2(8-6)\right]+12$

35. $4^2\left[(8-3)+6\right]$

36. $3^3\left[(6+5)-2^2\right]$

Objective 4   Know the meanings of $\neq, <, >, \leq,$ and $\geq$.

Tell whether each statement is **true** *or* **false.**

37. $95>97$

38. $41\leq 41$

39. $3\cdot 4\div 2^2\neq 3$

40. $3.25\geq 3.52$

41. $2\left[7(4)-3(5)\right]\geq 45$

42. $4\left[3\cdot 2+5(9)\right]\leq 105$

43. $\dfrac{2}{9}<\dfrac{1}{16}$

44. $4\frac{1}{4}\geq 3\frac{3}{4}$

45. $4\frac{1}{2}+2\frac{3}{4}<7$

46. $\dfrac{5+4\cdot 5}{14-2\cdot 3}\geq 2$

47. $4\geq \dfrac{2(3+1)-3(2+1)}{3\cdot 2-1}$

48. $\dfrac{5+2\cdot 3^2}{4^2-4\cdot 3}>1$

Objective 5   Translate word statements to symbols.

Write each word statement in symbols.

49. Seven equals thirteen minus five.

50. Nine is greater than sixteen.

51. Twelve is not equal to fourteen.

52. Twenty-two is greater than seventeen.

53. The sum of nine and thirteen is greater than twenty-one.

54. Nineteen is less than thirty-five.

55. Six is less than or equal to six.

56. The difference between thirty and seven is greater than twenty.

57. Seven is greater than the quotient of fifteen and five.

58. Seventeen is less than the product of three and ten.

Objective 6   **Write a statement that changes the direction of an inequality symbol.**

Write each statement with the inequality symbol reversed.

59. $9 < 12$

60. $12 \geq 8$

61. $\dfrac{1}{10} < 4$

62. $\dfrac{1}{2} < \dfrac{2}{3}$

63. $\dfrac{2}{3} < \dfrac{3}{4}$

64. $\dfrac{2}{7} \leq \dfrac{4}{5}$

65. $.21 > .19$

66. $.921 \leq .922$

67. $1 \geq 0$

68. $.1 > .01$

1.1   Mixed Exercises

Find the value of each expression.

69. $\left(\dfrac{5}{3}\right)^2$

70. $(4.5)^3$

71. $3 \cdot 4 + 5 \cdot 8$

72. $40 \div 5 + 3 \cdot 4$

73. $\dfrac{3 \cdot 8 - 4 \cdot 5}{2\left(3^2 - 1\right)}$

74. $2\left[4 + 2\left(5^2 - 3\right)\right]$

*Tell whether each statement is **true** or **false**.*

75. $5\left[3(4) - 7(6)\right] \leq 44$

76. $\dfrac{8 + 4 \cdot 5^2}{2 \cdot 3^2 + 4} \geq 5$

Write each word statement in symbols.

77. Five times the sum of two and nine is less than one hundred six.

78. Twenty is greater than or equal to the product of two and seven.

Write each statement with the inequality symbol reversed.

79. $.002 > .0002$

80. $\dfrac{3}{8} \leq \dfrac{3}{7}$

THE REAL NUMBER SYSTEM

1.2   Variables, Expressions, and Equations

Objective 1   Evaluate algebraic expressions, given values for the variables.

Find the value of each expression if $x = 2$ and $y = -4$.

1. $x + 3$

2. $8x^2 - 6x$

3. $2x^3 - y^2$

4. $x^4 + y^2$

5. $5x + 2 - 4y$

6. $2(7x - 3y)$

7. $\dfrac{x^2 + y}{x + 1}$

8. $\dfrac{2x + 3y}{3x - y + 2}$

9. $\dfrac{2x - 4}{2}$

10. $\dfrac{.2y}{.75}$

11. $\dfrac{3y^2 + 2x^2}{5x + y^2}$

12. $\dfrac{3x + y^2 + 2}{2x + 3y}$

Objective 2   Convert phrases from words to algebraic expressions.

Change the word phrases to algebraic expressions. Use x as the variable.

13. Four added to a number

14. Five times a number

15. A number subtracted from three

16. The product of four less than a number and two

17. The difference between twice a number and 7

18. Five subtracted from three times a number

19. Ten times a number, added to 21

20. The difference between four times a number and 15

21. The sum of a number and 4 is divided by twice the number

22. Half a number is subtracted from two-thirds of the number

Objective 3   Identify solutions of equations.

Determine whether the given number is a solution of the equation.

23. $x + 2 = 11;\ 9$

24. $x + 4 = 15;\ 11$

25. $6b + 2(b + 3) = 14;\ 2$

26. $5 + 3x^2 = 19;\ 2$

27. $\dfrac{p + 4}{p - 2} = 2;\ 6$

28. $\dfrac{m + 2}{3m - 10} = 1;\ 8$

29. $\dfrac{x^2 - 7}{x} = 6;\ 2$

30. $3y + 5(y - 5) = 22;\ 4$

31. $(a - 2)^3 = 27;\ 6$

32. $x^2 + 2x + 1 = 9;\ 3$

33. $7r + 5r - 8 = 16;\ 2$

34. $\dfrac{4-x}{x+2} = \dfrac{7}{5}; \dfrac{1}{2}$

Objective 4   Translate word statements to equations.

Change each word sentence to an equation. Use x as the variable.

35. The sum of a number and six is ten.

36. A number minus five is equal to nine.

37. The sum of five times a number and two is 23.

38. Five more than a number is 9.

39. The quotient of a number and nine is 17.

40. Four times a number is equal to two more than three times the number.

41. 10 divided by a number is two more than the number.

42. The product of six and a number is 18.

43. A number divided by two is zero.

44. The quotient of a number and ten is one.

Objective 5   Distinguish between expressions and equations.

Decide whether each of the following is an equation or an expression.

45. $3x+2y$

46. $8x=2y$

47. $9x+2y=2$

48. $4x+2y+7$

49. y^2-4y-3

50. $y^2-7y+4=0$

51. $\dfrac{x+4}{5}$

52. $\dfrac{x+3}{15}=2x$

53. $y=2x^2+4$

54. $5x^2+3$

1.2 Mixed Exercises

Find the value of each expression if $x=-2$ and $y=4$.

55. $9x-3y+2$

56. $.3(8x+2y)$

57. $\dfrac{4x-3y}{x+y+3}$

58. $\dfrac{3x}{4}-\dfrac{3y}{2}$

59. $\dfrac{4y^2-x^2}{2x^2+y}$

60. $\dfrac{5x+y^2}{-4x+2y}$

Decide whether the given number is a solution of the equation.

61. $2s^2 + 6 = 6;\ 2$

62. $\dfrac{2x-1}{x-2} = 3;\ 5$

63. $\dfrac{q^2+5}{q+15} = 7;\ 3$

64. $z^2 + 4z - 2 = 30;\ 4$

65. $(x+4)^2 = 29;\ 3$

66. $2(b+5) - 3b = 8;\ 2$

Change each word sentence to an equation. Use x as the variable.

67. The sum of a number and four is ten.

68. The product of six and 5 more than a number is nineteen.

69. Fourteen divided by a number is the sum of the number and two.

70. The quotient of twenty-four and a number is the difference between the number and two.

THE REAL NUMBER SYSTEM

1.3 Real Numbers and the Number Line

Objective 1 Classify numbers and graph them on the number line.

Use a real number to express each number in the following applications.

1. Carl withdrew $273 from his checking account. Then he deposited a check for $103.

2. During the past year, Abe made $3500 in the stock market.

3. Last year Nina lost 75 pounds.

4. The average number of pounds Nina lost in a week was 2.

5. Last year Lauren's savings account decreased by $72.

6. Mt. Whitney, one of the highest mountains in the United States, has an altitude of 14,495 feet.

7. Between 1970 and 1982, the population of Norway increased by 279,867.

8. It takes $2\frac{1}{2}$ hours longer to mow David's lawn than Faustino's.

9. When Stan graduated, he owed $42,500 on his college loans.

10. The Dead Sea, the saltiest body of water in the world, lies 396 meters below the level of the Mediterranean Sea.

Graph each group of rational numbers on a number line.

11. $-2, -1, 0, 2, 4$

12. $3, 5, -1, -3$

13. $6, 3, -3, -6, 0$

14. $-6, 3, -4, 1$

15. $\frac{1}{2}, 0, -3, -\frac{5}{2}$

16. $0, 2, 4, 6, \frac{17}{2}$

17. $-\frac{3}{2}, 0, \frac{3}{2}, \frac{7}{2}$

18. $-3\frac{1}{2}, -\frac{3}{2}, 0, \frac{1}{2}, 1$

19. $2\frac{1}{3}, 4, 6\frac{2}{3}, 8$

20. $-4.5, -2.3, 1.7, 3.5$

Objective 2 Tell which of two real numbers is less than the other.

Select the smaller number in each pair.

21. $-15, -12$

22. $0, -2$

23. $-.802, -.820$

24. $-5.99, -6.01$

25. $-2, 1$

26. $\frac{2}{3}, -\frac{1}{2}$

Decide whether each statement is **true** *or* **false.**

27. $-76 < 45$

28. $-5 > -5$

29. $-12 > -10$

30. $3 < -4$

Objective 3   Find the opposite of a real number.

Find the opposite of each number.

31. 23

32. -25

33. 4.5

34. $\frac{3}{8}$

35. $-\frac{5}{7}$

36. $2\frac{3}{7}$

37. 0

38. $-(-4)$

39. $-(-22)$

Objective 4   Find the absolute values of a real number.

Simplify by removing absolute value symbols.

40. $|-4|$

41. $|143|$

42. $|0|$

43. $-|95|$

44. $-|-25|$

45. $|16-14|$

46. $-|49-39|$

47. $|-7.52|$

48. $-|-.9|$

49. $\left|\frac{1}{2}+\frac{1}{3}\right|$

50. $\left|1\frac{1}{2}-2\frac{1}{4}\right|$

51. $-\left|2\frac{3}{8}-4\frac{3}{4}\right|$

Select the smaller number in each pair.

52. $|-8|, |2|$

53. $-|-2|, -|5|$

54. $-|-7|, -|10|$

1.3   Mixed Exercises

Use an integer to express each number in the following applications of numbers.

55. Last year, Faustino's salary increased by $2573.

56. Last year, the Yankees lost only 27 games.

57. Mt. Hood in Oregon is 11,235 feet high.

Select the smaller number in each pair.

58. $-800, -799$

59. $12.01, 12.001$

60. $\frac{7}{12}, -\frac{11}{12}$

61. $-\frac{2}{5}, -\frac{1}{4}$

62. $-(-4), -4$

63. $-|2|, -|-3|$

64. $|-12|, |13|$

65. $-|-2.5|, |2.4|$

THE REAL NUMBER SYSTEM

1.4 Adding Real Numbers

Objective 1 Add two numbers with the same sign.

Use a number line to find the sums.

1. $8+7$

2. $9+12$

3. $7+(-12)$

4. $-4+(-6)$

5. $-9+(-9)$

6. $-7+(-11)$

Objective 2 Add numbers with different signs.

Find the following sums.

7. $7+(-5)$

8. $-9+4$

9. $9+(-16)$

10. $-\dfrac{4}{5}+\dfrac{3}{5}$

11. $\dfrac{7}{12}+\left(-\dfrac{3}{4}\right)$

12. $-\dfrac{4}{7}+\dfrac{3}{5}$

13. $3\frac{5}{8}+\left(-2\frac{1}{4}\right)$

14. $14.1+(-14.1)$

15. $-10.475+6.325$

Objective 3 Add mentally.

*Perform each operation and then determine whether the statement is **true** or **false**. Try to do all work in your head.*

16. $19+(-13)=6$

17. $-14+11=-3$

18. $4.9+(-2.6)=-2.3$

19. $(-14)+15+(-2)=-3$

20. $\dfrac{3}{5}+\left(-\dfrac{3}{10}\right)=-\dfrac{3}{10}$

21. $-\dfrac{3}{8}+\dfrac{11}{12}=-\dfrac{7}{24}$

22. $5\frac{3}{8}+\left(-4\frac{1}{2}\right)=2\frac{1}{8}$

23. $-9\frac{1}{9}+8\frac{5}{6}=-\frac{5}{18}$

24. $\left|14+(-22)\right|=-14+22$

25. $\left|-5+(-4)\right|=5+4$

Objective 4   Use the order of operations with real numbers.

Find the following sums.

26. $6+\left[2+(-7)\right]$

27. $10+\left[4+(-27)\right]$

28. $-9+\left[5+(-19)\right]$

29. $-2+\left[-16+(-2)\right]$

30. $-14+3+\left[8+(-13)\right]$

31. $-2+\left[4+(-18+13)\right]$

32. $\left[(-7)+14\right]+\left[(-16)+3\right]$

33. $-7.6+\left[5.2\right]+(-11.4)$

34. $\dfrac{3}{8}+\left[-\dfrac{2}{3}+\left(-\dfrac{7}{12}\right)\right]$

35. $-\dfrac{4}{5}+\left[\dfrac{1}{4}+\left(-\dfrac{2}{3}\right)\right]$

36. $\left[\dfrac{7}{10}+\left(-\dfrac{3}{5}\right)\right]+\dfrac{1}{2}$

37. $\left[2\tfrac{1}{2}+\left(-3\tfrac{1}{4}\right)\right]+1\tfrac{1}{8}$

Objective 5   Translate words and phrases that indicate addition.

Write a numerical expression for each phrase, and then simplify the expression.

38. The sum of –9 and 14

39. 12 more than –7

40. –2 increased by 16

41. The sum of –8 and 3 and 2

42. 10 added to the sum of –4 and –3

43. 7 more than –2, increased by 9

44. 10 increased by the sum of –20 and 9

45. The sum of –8 and –4 and –11

46. The sum of –14 and –29, increased by 27

47. –10 added to the sum of 20 and –4

Solve each problem by writing a sum of real numbers and adding. No variables are needed.

48. A football team gained 4 yards from scrimmage on the first play, lost 21 yards on the second play, and gained 9 yards on the third play. How many yards did the team gain or lose altogether?

49. Pablo has $723 in his checking account. He write two checks, one for $358 and the other for $75. Finally, he deposits $205 in the account. How much does he now have in his account?

50. Charlie starts to climb a mountain at an altitude of 2324 feet. He climbs so that he gains 247 feet in altitude. Then he finds that, because of an obstruction, he must descend 15 feet. Then he climbs 98 feet up. What is his final altitude?

51. Penny owes $48 to a credit card company. She makes a purchase of $45 with her card, and then pays $77 to the company. How much does she still owe?

52. The temperature at dawn in Blackwood was 24°F. During the day the temperature decreased 30°. Then it increased 11° by sunset. What was the temperature at sunset?

53. A hot-air balloon rises 200 feet above the ground. It drops 50 feet, then drops another 25 feet, and finally rises 135 feet. How high is the balloon now?

1.4 Mixed Exercises

Find the following sums.

54. $-20+(-20)$

55. $25+(-26)$

56. $5+[-4+(-7)]$

57. $[(-6)+7]+(-8)$

58. $[9+(-21)]+[-17+(-6)]$

59. $-8.9+[6.8+(-4.7)]$

60. $[\frac{3}{5}+(-\frac{3}{10})]+\frac{2}{5}$

61. $[-2\frac{3}{8}+(-3\frac{1}{4})]+(-5\frac{3}{4})$

Write a numerical expression for each phrase, and then simplify the expression.

62. The sum of 4 and −12, increased by 6

63. −2 added to the sum of 25 and −20

64. The sum of −3 and −4 and −8

65. 20 more than the sum of −2 and −4

THE REAL NUMBER SYSTEM

1.5 Subtracting Real Numbers

Objective 1   Find a difference.

Use a number line to find the difference.

1. $10-5$

2. $7-10$

3. $4-4$

4. $-3-9$

5. $-7-2$

6. $3-9$

Objective 2   Use the definition of subtraction.

Find each difference.

7. $14-20$

8. $32-36$

9. $-9-4$

10. $-4-7$

11. $13-(-9)$

12. $1-(-7)$

13. $-4-(-20)$

14. $-3.2-(-7.6)$

15. $4.5-(-2.8)$

16. $\frac{1}{2}-\left(-\frac{1}{10}\right)$

17. $-\frac{3}{10}-\left(-\frac{4}{15}\right)$

18. $3\frac{3}{4}-\left(-2\frac{1}{8}\right)$

Objective 3   Work subtraction problems that involve brackets.

Work each problem.

19. $4-\left[6+(-9)\right]$

20. $-.2-\left[.6+(-.9)\right]$

21. $\left[6-(-14)\right]-26$

22. $\left[8-(-7)\right]-2$

23. $\left[3+(-9)\right]-(-6)$

24. $3-\left[-4+(11-19)\right]$

25. $\left[\frac{1}{3}-\left(-\frac{1}{5}\right)\right]-\left(-\frac{4}{15}\right)$

26. $\frac{2}{9}-\left[\frac{5}{6}-\left(-\frac{2}{3}\right)\right]$

27. $-4+\left[(-5-3)-(-2+4)\right]$

28. $-2+\left[(-12+10)-(-4+2)\right]$

| Objective 4 | Translate words and phrases that indicate subtraction. |

Write a numerical expression for each phrase, and then simplify the expression.

29. The difference between –9 and 3

30. The difference between –4 and –13

31. The difference between –6 and –2

32. 4 less than the difference between –7 and –9

33. 4 less than -4

34. 7 less than -4

35. –8 decreased by 2 less than –1

36. –4 decreased by 1 less than –4

37. –6 subtracted from the sum of 4 and –7

38. –12 subtracted from the sum of –4 and –2

Solve each of the following problems by writing a difference between real numbers and subtracting. No variables are needed.

39. Dr. Somers runs an experiment at –43.3°C. He then lowers the temperature by 7.9°C. What is the new temperature for the experiment?

40. The highest point in the state of Washington, Mt. Rainier, has an elevation of 14,410 feet. The highest point in Oregon, Mt. Hood, has an elevation of 11,235 feet. How much taller is Mt. Rainier than Mt. Hood?

41. David has a checking account balance of $439.42. He overdraws his account by writing a check for $702.58. Write his new balance as a negative number.

42. At 1:00 A.M., the temperature on the top of Mt. Washington in New Hampshire was –12°F. At 11:00 A.M., the temperature was 25°F. What was the rise in temperature?

43. The highest point in a country has an elevation of 1408 meters. The lowest point is 396 meters below sea level. Using zero as sea level, find the difference between the two elevations.

44. Carla owes $917.34 on her VISA account. She makes additional purchases which total $229.98. Express her new balance as a negative number.

1.5 Mixed Exercises

Work each problem.

45. $45 - 55$

46. $-7 - (-14)$

47. $22 - (-24)$

48. $-(-25) - 45$

49. $-5.6 - (-5.6)$

50. $-7.2 - 8.9$

51. $6 - \left[-8 - (-4) \right]$

52. $\left[8 - (-12) \right] - 2$

53. $\left(\frac{1}{2} - \frac{1}{3} \right) - \frac{5}{6}$

54. $\left[\frac{5}{8} - \left(-\frac{1}{16} \right) \right] - \left(-\frac{3}{8} \right)$

Write a numerical expression for each phrase, and simplify the expression.

55. The difference between 4 and −10

56. The sum of −4 and 12, decreased by 9

57. 2 less than the difference between 10 and −4

58. The sum of −4 and −8, decreased by 2

THE REAL NUMBER SYSTEM

1.6 Multiplying and Dividing Real Numbers

Objective 1   Find the product of numbers with different signs.

Find the products.

1. $7(-4)$

2. $(-10)(12)$

3. $(-12)(6)$

4. $12(-11)$

5. $(-80)(4)$

6. $11(-4)$

7. $\left(\frac{1}{5}\right)\left(-\frac{2}{3}\right)$

8. $\left(-\frac{3}{8}\right)\left(\frac{14}{9}\right)$

9. $\left(-\frac{2}{7}\right)\left(\frac{21}{26}\right)$

10. $7(-2.5)$

11. $(7.2)(-5)$

12. $(-3.2)(4.1)$

Objective 2   Find the product of two negative numbers.

Find the products.

13. $(-3)(-4)$

14. $(-7)(-2)$

15. $(-13)(-14)$

16. $(-10)(-100)$

17. $(-17)(-17)$

18. $(-4)(-10)$

19. $\left(-\frac{2}{7}\right)\left(-\frac{14}{5}\right)$

20. $\left(-\frac{1}{3}\right)\left(-\frac{9}{4}\right)$

21. $\left(-\frac{3}{10}\right)\left(-\frac{5}{9}\right)$

22. $(-1.3)(-2.1)$

23. $(-7.1)(-.4)$

24. $(-.4)(-3.4)$

Objective 3   Use the reciprocal of a number to apply the definition of division.

Find the reciprocal, if one exists, for each number.

25. 4

26. -2

27. $-\frac{1}{4}$

28. 0

29. $\frac{3}{5}$

30. $-\frac{1}{3}$

31. $-\frac{11}{12}$

32. $4\frac{1}{24}$

33. $.25$

34. $-.125$

35. $10+(-10)$

36. $5\frac{3}{8}$

Find the quotients.

37. $\frac{40}{-5}$

38. $\frac{-72}{9}$

39. $\frac{44}{-11}$

40. $\frac{-120}{-20}$

41. $\frac{10}{0}$

42. $\frac{0}{-2}$

43. $-\frac{3}{16} \div \frac{9}{8}$

44. $-\frac{1}{3} \div \left(-\frac{8}{9}\right)$

45. $-\frac{27}{35} \div \left(-\frac{9}{5}\right)$

46. $\frac{14}{33} \div \left(-\frac{7}{11} \right)$ **47.** $-20.8 \div (-4)$ **48.** $(-5.5) \div 2.2$

Objective 4   Use the order of operations when multiplying and dividing signed numbers.

Perform the indicated operations.

49. $24 - 5 \cdot 7$

50. $7 \cdot 4 - 2 \cdot 12$

51. $4(-8) - (-2)(-7)$

52. $(-4)(9) - (-5)(4)$

53. $-3(4) - 4(-2)$

54. $-9(7-4)$

55. $-4[(-2)(7) - 2]$

56. $-7[-4 - (-2)(-3)]$

Simplify the numerators and denominators separately. Then find the quotients.

57. $\dfrac{9(-4)}{-6 - (-2)}$

58. $\dfrac{4(7)}{12 + (-28)}$

59. $\dfrac{8(4)}{6(4) + 4(-2)}$

60. $\dfrac{-3 - (-4 + 1)}{-7 - (-6)}$

61. $\dfrac{5(-8 + 3)}{13(-2) + (-6 - 1)(-4 + 1)}$

62. $\dfrac{6^2 - 4}{2(2) + 4(-2)}$

63. $\dfrac{4(-6) - (3)(8)}{2(-7) + (-2)(-11)}$

64. $\dfrac{-4[8 - (-3 + 7)]}{-6[3 - (-2)] - 3(-3)}$

65. $\dfrac{9(-2 + 4)}{15(-3) + (-7 - 4)(-9 + 6)}$

66. $\dfrac{2^2 + 4^2}{5^2 - 3^2}$

67. $\dfrac{4^3 - 3^3}{-5(-4 + 2)}$

68. $\dfrac{5^2 + 4^2}{2^2 + 3}$

Objective 5   Evaluate expressions involving variables.

Evaluate the following expressions if $x = -3,\ y = 2,$ *and* $a = 4$.

69. $-x + 2y - 3a$

70. $-3x + 4y - (a - x)$

71. $(y - 2x)(-3a)$

72. $-x^2 + 3y$

73. $(2y + 4)(a) + |y|$

74. $(x - 2)(4 - y)$

75. $(x - y) - (a - 3y)$

76. $(-2y + 4a) - (3x + y)$

77. $2(x - 3)^2 + 2y^2$

78. $\dfrac{2x^2 - 3y}{4a}$

81. $\dfrac{4a - x}{y^2}$

79. $\dfrac{x^2 - y^2}{a^2}$

82. $\dfrac{-x^2 + 2y}{a}$

80. $\dfrac{3a^2 + x}{2y}$

Objective 6   Translate words and phrases that indicate multiplication and division.

Write a numerical expression for each phrase and simplify.

83. The product of 7 and –2, added to 4

84. The product of –7 and 3, added to –7

85. The product of 10 and –2, subtracted from –2

86. The difference between –12 and the product of –2 and 7

87. Twice the sum of 14 and –4, added to –2

88. Three-tenths of the difference between 50 and –10, subtracted from 85

89. 70% of the sum of 20 and –4

90. 85% of the difference between 32 and –4

91. –34 subtracted from two-thirds of the sum of 16 and –10

92. –7 added to seven-eighths of the difference between –2 and 6

93. The quotient of –108 and –4

94. The quotient of 50 and the sum of 35 and –5

95. The sum of –12 and the quotient of 49 and –7

96. The difference between 9 and the quotient of –12 and 4

97. The difference between –3 and the quotient of –12 and –4

98. The product of 40 and –3, divided by the difference between 5 and –10

99. The product of –4 and 7, divided by the sum of –3 and 14

100. The quotient of the sum of 14 and –4 and the difference between –11 and –9

Objective 7   Translate simple sentences into equations.

Write each statement in symbols, using x as the variable.

101. 5 times a number is –45.

102. The quotient of a number and –2 is –9.

103. The sum of a number and 9 is –8.

104. The difference between a number and –7 is 12.

105. Two-thirds of a number is –7.

106. The product of a number and –1 is 7.

107. When a number is divided by –4, the result is 1.

108. 9 less than a number is –4.

109. –8 times a number is 72.

1.6 Mixed Exercises

Perform the indicated operations.

110. $4(-12)$

111. $(-10)(-14)$

112. $-3(-5-9)$

113. $-4\left[-3-5(-2)\right]$

114. $(4-2)(-7-1)$

115. $(-7-2)(-2)-4$

116. $(-3-6)(-5)-10$

117. $\left|-4(9)\right|-\left|-11\right|$

Evaluate the following expression if $x = -1$, $y = 2$, **and** $a = -3$.

118. $-x^2 + 2a^2$

119. $(-4+x)(-a)-|x|$

120. $(x-4)(3-a)$

121. $(4x+3y)(2a-x)$

122. $3y^2 - 4x^2 + a^2$

123. $(x-2)^2 + 2x^3$

Perform the indicated operations.

124. $\dfrac{-9}{-9}$

125. $\dfrac{-120}{5}$

126. $\dfrac{0}{-4}$

127. $\dfrac{9}{0}$

128. $\dfrac{1}{2} \div \left(-\dfrac{1}{2}\right)$

129. $\dfrac{-7(2)-(-3)}{5+(-3)}$

130. $\dfrac{4(-10+3)}{9(-2)-3(-2)}$

131. $\dfrac{8^2 - 12}{5(-5)+3(4)}$

132. $\dfrac{2(-3)-(-6)(-7)}{-3(6)-(5)(1)}$

133. $\dfrac{-4\left[6-(-3+1)\right]}{-3\left[2-(-4)\right]-9(-2)}$

134. $\dfrac{-8\left[9-(-9+2)\right]}{14(-3)+(-9-8)(-5+3)}$

Write a numerical expression for each phrase and simplify the expression.

135. The quotient of 100 and the sum of –16 and –9

136. The product of –40 and 4, divided by the difference between 7 and –3

137. The quotient of the sum of –20 and –10 and the difference between 3 and –3

138. The sum of the quotient of –20 and –5 and the quotient of 100 and –25.

THE REAL NUMBER SYSTEM

1.7 Properties of Real Numbers

Objective 1   Use the commutative properties.

Complete each statement. Use a commutative property.

1. $y + 4 = $ _____ $+ y$

2. $5(2) = $ _____ (5)

3. $(ab)(2) = (2)$ _____

4. $7m = $ _____ (7)

5. $-4(p + 9) = $ _____ (-4)

6. $10\left(\frac{1}{4} \cdot 2\right) = $ _____ (10)

7. $-4\left(\frac{1}{5}\right) = \left(\frac{1}{5}\right)$ _____

8. $3 + (-4) = -4 + $ _____

9. $2 + \left[10 + (-9)\right] = $ _____ $+ 2$

10. $-4(4 + z) = $ _____ (-4)

Objective 2   Use the associative properties.

Complete each statement. Use an associative property.

11. $x(9y) = $ _____ (y)

12. $(4 \cdot 5)(-7) = $ _____ $\left[5(-7)\right]$

13. $\left[-4 + (-2)\right] + y = $ _____ $+ (-2 + y)$

14. $(2m)(-7) = (2)$ _____

15. $4(ab) = $ _____ $\cdot b$

16. $(-6x)(-2) = (-6)$ _____

17. $(-12x)(-y) = (-12)$ _____

18. $(-r)\left[(-p)(-1)\right] = $ _____ $(-q)$

19. $\left[x + (-4)\right] + 3y = x + $ _____

20. $4r + (3s + 14t) = $ _____ $+ 14t$

Objective 3   Use the identity property.

Simplify.

21. $4 + 0 = $

22. $-7 + 0 = $

23. $1(-4) = $

24. $7(1) = $

25. $\frac{30}{35} = $

Objective 4   **Use the inverse properties.**

Complete the statements so that they are examples of either an identity property or an inverse property. Identify which property is used.

26. $-4+$ _____ $=0$

27. _____ $+\frac{1}{7}=0$

28. $1\cdot$ _____ $=1$

29. $\frac{2}{7}\cdot$ _____ $=1$

30. $-\frac{3}{5}\cdot$ _____ $=1$

31. _____ $\cdot\frac{5}{8}=1$

32. $-14+$ _____ $=0$

33. $-9+$ _____ $=0$

34. _____ $+0=0$

35. _____ $\cdot-2\frac{5}{6}=1$

36. $.25+$ _____ $=0$

Objective 5   **Use the distributive property.**

Use the distributive property to rewrite each expression. In Exercises 37 and 38, simplify the result.

37. $6y+7y$

38. $10r-4r$

39. $a(z+2)$

40. $4\cdot r+4\cdot p$

41. $3(a+b)$

42. $4c-4d$

43. $n(2a-4b+6c)$

44. $7(a-4b)$

45. $-2(5y-9z)$

46. $-(-2k+7)$

47. $-14x+(-14y)$

48. $2(7x)+2(8z)$

1.7 Mixed Exercises

Label each statement as an example of the commutative, associative, identity, inverse, or distributive property.

49. $(13\cdot4)\cdot8=13\cdot(4\cdot8)$

50. $(9+12)+10=(12+9)+10$

51. $0+(-12)=-12$

52. $\left(-4\frac{1}{3}\right)+4\frac{1}{3}=0$

53. $(-8)\left[5(-2)\right]=\left[-8(5)\right](-2)$

54. $5(2x+4y)=5(2x)+5(4y)$

55. $\left(-\frac{4}{5}\right)+0=-\frac{4}{5}$

56. $1\left(-\frac{5}{6}\right)=-\frac{5}{6}$

57. $2(3m)+2(7n)=2(3m+7n)$

58. $(6\cdot2)(-5)=(2\cdot6)(-5)$

Use the properties of this section to simplify the following expressions.

59. $-4+4y+13$

60. $4r-2-8r+5$

61. $\left(\frac{2}{3}\right)(-11)\left(-\frac{3}{2}\right)$

62. $(-100)+(-42+100)$

63. $-(3x-2)$

64. $-(-4b-8)$

THE REAL NUMBER SYSTEM

1.8 Simplifying Expressions

Objective 1   Simplify Expressions.

Simplify each expression.

1. $14 + 3y - 8$

2. $3(4x - 2y)$

3. $4(2x + 5) + 7$

4. $-2 + 4(2n - 3)$

5. $-(9 - 4b) - 8$

6. $11 - (d - 2) + (-6)$

7. $7 - 6y + (5 - 2)$

8. $-4 + s - (12 - 21)$

9. $-2(-5x + 2) + 7$

10. $4(-6p - 2) + 2 - 4$

Objective 2   Identify terms and numerical coefficients.

Give the numerical coefficient of each term.

11. $4x$

12. $-2y^2$

13. $-7a^2$

14. $4s^4$

15. $.3a^2b$

16. $-12xy$

17. 125

18. z^5

19. -4^2s^4

20. $-\frac{3}{5}a^2b$

21. $-\frac{5}{9}v^6w^4$

22. $\dfrac{7x}{9}$

Objective 3   Identify like terms.

Identify each group of terms as like *or* unlike.

23. $2x, 7x$

24. $9, -2a$

25. $-7q^2, 2q^2$

26. $-8m, -8m^2$

27. $2w, 4w, -w$

28. $4k, -9k, 3k$

29. $-5y, -4y, 2$

30. $4x^2, -7x^2$

31. $4x, -10x^2, -9x^2$

32. $18, 18y$

33. $2, -4, 16$

34. $7xy, -6xy^2$

Objective 4   **Combine like terms.**

Simplify each expression by combining like terms.

35. $23a - 16a$

36. $12 - 4x - 2 - 7x$

37. $2.3r + 6.9 + 2.8 + 3.6r$

38. $4a^2 - 4a^3 - 2a^2 + 7a^3$

39. $\frac{1}{3} + \frac{3}{4}y - \frac{5}{6} - \frac{2}{3}y$

40. $\frac{7}{10}r + \frac{3}{10}s - \frac{2}{5}r - \frac{4}{5}s$

Use the distributive property and combine like terms to simplify the following expressions.

41. $2(3x+5)$

42. $-4(2-3x)$

43. $7r - (2r+4)$

44. $2(-3+t) - 4t$

45. $4(2q+7) - (3q-2)$

46. $-6(a+2) + 4(2a-1)$

47. $-5(s+4) + 4(2s+2)$

48. $2(4x-1) - (5x+2)$

49. $2.5(3y+1) - 4.5(2y-3)$

50. $-.8(7t-5) + .6(9t-7)$

Objective 5   **Simplify expressions from word phrases.**

Write each phrase as a mathematical expression and simplify by combining like terms. Use x as the variable.

51. Seven times a number, added to twice the number

52. Four times a number, subtracted from three times the number

53. The sum of six times a number and 12, added to four times the number

54. Three times the sum of 9 and twice a number, added to four times the number

55. The sum of seven times a number and 2, subtracted from three times the number

56. The difference between five times a number and 3, added to four times the sum of the number and 2

57. The sum of ten times a number and 7, subtracted from the difference between 2 and nine times the number

58. Twelve times the difference between 4 and twice a number, subtracted from 10

59. Four times the difference between twice a number and six times the number, added to six times the sum of the number and 9

60. Four times the difference between twice a number and –10, subtracted from three times the sum of –7 and five times the number

1.8 Mixed Exercises

Simplify the following expressions.

61. $6(3a-2b)$

62. $3+2(7x-4)$

63. $7(5n-2)-(6-11)$

64. $12y-7y^2+4y-3y^2$

65. $.8y^2-.2xy-.3xy+.9y^2$

66. $-4(x+4)+2(3x+1)$

67. $-5(s-6)+4(2s+3)$

68. $\frac{1}{2}(2x-4)-\frac{3}{4}(8x+12)$

Give the numerical coefficient of each term.

69. $93ab^2c$

70. $-9x^2y^2$

71. $\frac{1}{10}ab$

72. $5.6r^5$

Identify each group of terms as **like** *or* **unlike.**

73. $4ab, 12ab$

74. $5z^3, 5z^2, 5z^2$

75. $\frac{1}{3}, -\frac{3}{4}, 4$

Chapter 2

SOLVING EQUATIONS AND INEQUALITIES

2.1 The Addition Property of Equality

Objective 1 **Identify linear equations.**

Tell whether each of the following is a linear equation.

1. $9x + 2 = 0$

2. $3x^2 + 4x + 3 = 0$

3. $7x^2 = 10$

4. $3x^3 = 2x^2 + 5x$

5. $\frac{5}{x} - \frac{3}{2} = 0$

6. $4x - 2 = 12x + 9$

Objective 2 **Use the addition property of equality.**

Solve each equation by using the addition property of equality. Check each solution.

7. $y - 4 = 16$

8. $r + 9 = 8$

9. $3x + 2 = 5x + 12$

10. $3y = 7y - 4$

11. $p - \frac{2}{3} = \frac{5}{6}$

12. $y + 4\frac{1}{2} = 3\frac{3}{4}$

13. $\frac{2}{3}t - 5 = \frac{5}{3}t$

14. $\frac{9}{8}p - \frac{1}{2} = \frac{1}{8}p$

15. $5.7x + 12.8 = 4.7x$

16. $9.5y - 2.4 = 10.5y$

Objective 3 **Simplify equations, and then use the addition property of equality.**

Solve each equation. First simplify each side of the equation as much as possible. Check each solution.

17. $6x - 3x + 10 = -2$

18. $6x + 3x - 7x + 4 = 10$

19. $3(t+3) - (2t+7) = 9$

20. $5x + 4(2x+1) - (5x-1-2) = 9$

21. $-4(5g-7) + 3(8g-3) = 15 - 4 + 3g$

22. $10x + 4x - 11x + 4 - 7 = 2 - 4x - 3 + 8x$

23. $4(3a-2) - 6(2+a) = 5(2a-5)$

24. $2(4t+6) - 3(2t-3) = -3(3t-4) + 5 - t$

25. $-7(1+2b) - 6(3-5b) = 5(4+3b) - 45$

26. $8(2-4b) + 3(5-b) = 4(1-9b) + 22$

27. $\frac{8}{5}t + \frac{1}{3} = \frac{5}{6} + \frac{3}{5}t - \frac{1}{6}$

28. $\frac{5}{12} + \frac{7}{6}s - \frac{1}{6} = \frac{5}{6}s + \frac{1}{4} - \frac{2}{3}s$

29. $3.6p + 4.8 + 4.0p = 8.6p - 3.1 + .7$ **30.** $.03x + 0.6 + .09x - .9 = 2.1$

2.1 Mixed Exercises

Solve each equation.

31. $2z + 8 = -12$ **32.** $4x + 11 = 7x + 15$

33. $19k = 18k - 7$ **34.** $-7t + 12 = -4t$

35. $r - \frac{4}{9} = \frac{1}{12}$ **36.** $r - \frac{5}{8} = \frac{3}{4}$

37. $0.04x + 0.4 = 0.06x + 9.4$

SOLVING EQUATIONS AND INEQUALITIES

2.2 The Multiplication Property of Equality

Objective 1 Use the multiplication property of equality.

Solve each equation and check your solution.

1. $8x = 24$

2. $7y = 63$

3. $-3w = 42$

4. $-16a = -48$

5. $\frac{b}{5} = 4$

6. $\frac{u}{2} = 3$

7. $\frac{3p}{7} = -6$

8. $\frac{b}{-2} = 21$

9. $-9k = 81$

10. $-5m = -35$

11. $\frac{3}{4}r = -27$

12. $-\frac{7}{2}t = -4$

13. $\frac{y}{4} = \frac{1}{3}$

14. $\frac{6}{7}y = \frac{2}{3}$

15. $.9x = 5.4$

16. $4.9q = 24.5$

17. $2.1a = 9.03$

18. $3.5b = 24.85$

19. $1.9k = 11.02$

20. $.81m = 2.916$

21. $7.5p = -61.5$

22. $6.3x = -15.12$

23. $-2.7v = -17.28$

24. $-5.3w = 8.48$

25. $4.3r = -11.61$

26. $-5.9y = -21.24$

Objective 2 Simplify equations, and then use the multiplication property of equality.

Solve each equation and check your solution.

27. $4r + 3r = 63$

28. $3x + 6x = 72$

29. $7y - 2y = 45$

30. $9z - 3z = 24$

31. $10a - 7a = -24$

32. $14m - 6m = -56$

33. $4v + 3v + 7v = 98$

34. $8f + 4f - 3f = 72$

35. $8s - 3s + 4s = 90$

36. $-y = 3.9$

37. $-q = -4$

38. $-t = -26$

39. $-h = \frac{7}{4}$

40. $3b - 4b = 8$

41. $9p - 10p = -18$

42. $3w - 7w = 20$

43. $7q - 10q = -24$

44. $4x - 8x + 2x = 16$

45. $2f + 3f - 7f = 48$

46. $-11h - 6h + 14h = -21$

2.2   Mixed Exercises

Solve each equation and check your solution.

47. $9y = -72$

48. $\frac{y}{7} = -15$

49. $\frac{4}{3}z = 8$

50. $\frac{3}{8}z = -\frac{5}{4}$

51. $1.4c = -2.1$

52. $-3.5q = 33.25$

53. $12r + 3r = -90$

54. $-7b + 12b = 125$

55. $-f = 26$

56. $-t = -9.5$

57. $18r - 6r + 3r = -105$

58. $17x + 9x - 11x = -9$

SOLVING EQUATIONS AND INEQUALITIES

2.3 More on Solving Linear Equations

Objective 1   Learn the four steps for solving a linear equation and apply them.

Solve each equation and check your solution.

1. $7t + 6 = 11t - 4$

2. $7x + 11 = 9x + 25$

3. $7j + 1 = 10j - 29$

4. $4 + x = -(x + 6)$

5. $4(z - 2) - (3z - 1) = 2z - 6$

6. $3(x + 4) = 6 - 2(x - 8)$

7. $3 - (1 - y) = 3 + 5y$

8. $-(v + 2) = 3 + v$

9. $4w - 5w + 3(w - 7) = -4(w + 4) + 7$

10. $3a - 6a + 4(a - 4) = -2(a + 2)$

11. $4r - 3(3r - 2) = 8 - 3(r - 4)$

12. $3(t + 5) = 6 - 2(t - 4)$

13. $6f - 8f + 4(f - 3) = -2(f + 4)$

14. $4r - 8r + 3(r - 4) = -2(r + 5)$

Objective 2   Solve equations with fractions or decimals as coefficients.

Solve each equation.

15. $\frac{1}{5}(z - 5) = \frac{1}{3}(z + 2)$

16. $\frac{3}{8}x - \frac{1}{3}x = \frac{1}{12}$

17. $\frac{2}{3}y - \frac{1}{4}y = -\frac{5}{12}y + \frac{1}{2}$

18. $\frac{1}{3}(2m - 1) - \frac{3}{4}m = \frac{5}{6}$

19. $\frac{5}{6}(r - 2) - \frac{2}{9}(r + 4) = \frac{7}{18}$

20. $\frac{1}{2}r + \frac{5}{14}r = r - \frac{4}{7}$

21. $\frac{1}{8}(t - 3) + \frac{3}{8}(t + 2) = t - 2$

22. $\frac{3}{8}x - \left(x - \frac{3}{4}\right) = \frac{5}{8}(x + 3)$

23. $.90x = .40(30) + .15(100)$

24. $.35(20) + .45y = .125(200)$

25. $.12x + .24(x - 5) = .56x$

26. $.76t + .80(11 - t) = .45(20)$

27. $.24x - .38(x + 2) = -.34(x + 4)$

28. $.45a - .35(20 - a) = .02(50)$

29. $.07(10,000)+.02x=.03(10,000+x)$ **30.** $.01r+.325(1000)=.05(1000+r)$

Objective 3   Solve equations that have no solution or infinitely many solutions.

Solve each equation.

31. $3(6x-7)=2(9x-6)$

32. $4x+15=4(x-6)$

33. $6y-3(y+2)=3(y-2)$

34. $-1-(2+y)=-(-4+y)$

35. $8k+14=2(k+2)+3(2k+1)$

36. $4(2p-3)-3(3p+1)=-18-p+3$

37. $6(6t+1)=9(4t-3)+11$

38. $7y-11=6(2y+3)-5y$

39. $4b-2b-16-b=4b+16-3b$

40. $3(r-2)-r+4=2r+6$

41. $8(2d-4)-3(7d+8)=-5(d+2)$ **42.** $4(4-4h)+2(3+3h)=12+5(2-2h)$

43. $2(5w-3)+11=3(3w+1)+5(w-3)-4w$

44. $7(3b-4)-4(2b+2)=4-2(b-3)+5(3b-6)$

Objective 4   Write expressions for two related unknown quantities.

Write an expression for the two related unknown quantities.

45. Two numbers have a sum of 36. One is *m*. Find the other number.

46. The product of two numbers is 17. One number is *p*. What is the other number?

47. A cashier has *q* dimes. Find the value of the dimes in cents.

2.3 Mixed Exercises

Solve each equation.

48. $2x-5x-4=11$

49. $-7z+2z-2+3z=8$

50. $5y-4y+6y-9=2y+6$

51. $2z+8+5z+4=9z+7$

52. $10+3(b+2)=16+5b$

53. $3x-5(x+2)=11+(x-9)$

54. $4y+2(2y+1)-(5y-1-5)=9$

55. $-(5+m)-3(m+2)=6-3$

56. $-(4x+2)-(-3x-5)=3$

57. $3(p-4)+4p=8+p+1-p$

58. $2(2y-5)=4y+10$

59. $5(3r-4)=4(2r+2)+7r$

60. $-2(4x+7)=-4(2x+4)+2$

61. $4(2.2s+5)=7.8s-15$

62. $4(4y-5)+6=2(5y+2)+3y$

63. $5(4z+2)+5=2(10z+6)+3$

64. $-5(4a-2)+2(3a-7)=-4(4a+1)-3$

Write an expression for the two related unknown quantities.

65. Admission to the circus costs x dollars for an adult and y dollars for a child. Find the total cost of 6 adults and 4 children.

SOLVING EQUATIONS AND INEQUALITIES

2.4   An Introduction to Applications of Linear Equations

Objective 1   Learn the six steps for solving applied problems.

Objective 2   Solve problems involving unknown numbers.

Objective 3   Solve problems involving sums of quantities.

Write an equation for each of the following and then solve the problem. Use x as the variable.

1.   If 4 is added to 3 times a number, the result is 7. Find the number.

2.   If 2 is subtracted from four times a number, the result is 3 more than six times the number. What is the number?

3.   If −2 is multiplied by the difference between 4 and a number, the result is 24. Find the number.

4.   Six times the difference between a number and 4 equals the product of the number and −2. Find the number.

5.   When the difference between a number and 4 is multiplied by −3, the result is two more than −5 times the number. Find the number.

6.   If four times a number is added to 7, the result is five less than six times the number. Find the number.

Solve each problem.

7.   A rope 116 inches long is cut into three pieces. The middle-sized piece is 10 inches shorter than twice the shortest piece. The longest piece is $\frac{5}{3}$ as long as the shortest piece. What is the length of the shortest piece?

8.   George and Al were opposing candidates in the school board election. George received 21 more votes than Al, with 439 votes cast. How many votes did Al receive?

9.   On a psychology test, the highest grade was 38 points more than the lowest grade. The sum of the two grades was 142. Find the lowest grade.

10.   Mount McKinley is Alaska is 5910 feet higher than Mount Rainier in Washington. Together, their heights total 34,730 feet. How high is each mountain?

11. Charles bought five general admission tickets and four student tickets for a movie. He paid $35.25. If each student ticket cost $3.50, how much did each general admission ticket cost?

12. Penny is making punch for a party. The recipe requires twice as much orange juice as cranberry juice and 8 times as much ginger ale as cranberry juice. If she plans to make 176 ounces of punch, how much of each ingredient should she use?

13. Pablo, Faustino, and Mark swim at a public pool each day for exercise. One day Pablo swam five more than three times as many laps as Mark, and Faustino swam four times as many laps as Mark. If the men swam 29 laps altogether, how many laps did each one swim?

14. Linda wishes to build a rectangular dog pen using 52 feet of fence and the back of her house, which is 36 feet long to enclose the pen. How wide will the dog pen be if the pen is 36 feet long?

Objective 4   **Solve problems involving supplementary and complementary angles.**

Solve each problem.

15. Find the measure of an angle if the measure of the angle is 8° less than three times the measure of its supplement.

16. Find the measure of an angle whose supplement measures 20° more than twice its complement.

17. Find the measure of an angle such that the sum of the measures of its complement and its supplement is 138°.

18. Find the measure of an angle such that the difference between the measure of its supplement and twice the measure of its complement is 49°.

19. Find the measure of an angle whose complement is 9° more than twice its measure.

20. Find the measure of an angle whose supplement measures 3 times its complement.

21. Find the measure of an angle whose supplement measures 6° more than 7 times its complement.

22. Find the measure of an angle such that the difference between the measures of an angle and its complement is 20°.

23. Find the measure of an angle if its supplement measures 15° less than four times its complement.

24. Find the measure of an angle if its supplement measures 4° less than three times its complement.

Objective 5 Solve problems involving consecutive integers.

Solve each problem.

25. Find two consecutive even integers whose sum is 154.

26. Find two consecutive even integers such that the smaller, added to twice the larger, is 292.

27. Find two consecutive integers such that the larger, added to three times the smaller, is 109.

28. Find two consecutive odd integers such that if three times the smaller is added to twice the larger, the sum is 69.

29. Find two consecutive odd integers such that the larger, added to eight times the smaller, equals 119.

2.4 Mixed Exercises

Solve each problem.

30. If a number is subtracted from 83, the result is 19 more than 37. Find the number.

31. Ron runs a ski train. One day he noticed that the train contained 13 more women than men (including himself). If there were a total of 165 people on the train, how many of them were men?

32. A piece of rope is 130 centimeters long. It is cut into three pieces. The longest piece is 6 less than 3 times as long as the shortest piece, and the middle-sized piece is 26 centimeters longer than the shortest piece. Find the lengths of the three pieces.

33. Find the measure of an angle such that the difference between the measure of the angle and the measure of its complement is 28°.

34. Find three consecutive odd integers whose sum is 363.

35. Find the measure of an angle whose complement measures 30° less than half its supplement.

36. When the sum of a number and 9 is multiplied by 3, the result is 90. Find the number.

37. Find three consecutive integers such that the sum of the first two is 74 more than the third.

38. Find the measure of an angle such that the sum of the measures of its supplement and its complement is 114°.

39. Marge invested $240 more in stock A than in stock B. If she invested a total of $600 in the two stocks, how much did she invest in stock A?

SOLVING EQUATIONS AND INEQUALITIES

2.5 Formulas and Applications from Geometry

Objective 1   Solve a formula for one variable given the values of the other variables.

In the following exercises, a formula is given, along with the values of all but one of the variables in the formula. Find the value of the variable that is not given.

1. $V = LWH$; $L = 2$, $W = 4$, $H = 3$ 2. $P = 2L + 2W$; $P = 42$; $W = 6$

3. $A = \frac{1}{2}bh$; $b = 8$, $h = 2.5$ 4. $V = \frac{1}{3}Bh$; $B = 27$, $V = 63$

5. $C = 2\pi r$; $C = 43.96$, $\pi = 3.14$ 6. $A = \pi r^2$; $r = 3$, $\pi = 3.14$

7. $I = prt$; $I = 288$, $r = .04$, $t = 3$ 8. $C = \frac{5}{9}(F - 32)$; $F = 104$

9. $F = \frac{9}{5}C + 32$; $C = 35$ 10. $A = \frac{1}{2}(b + B)h$; $b = 6$, $B = 16$, $A = 132$

11. $V = \frac{4}{3}\pi r^3$; $r = 3$, $\pi = 3.14$ 12. $V = \frac{1}{3}\pi r^2 h$; $r = 4$, $h = 6$, $\pi = 3.14$

Objective 2   Use a formula to solve an applied problem.

Use a formula to write an equation for each of the following applications; then solve the application.

13. Find the length of a rectangular garden if its perimeter is 96 feet and its width is 12 feet.

14. Find the height of a triangular banner whose area is 48 square inches and base is 12 inches.

15. Ruth has 42 feet of binding for a rectangular rug that she is weaving. If the rug is 9 feet wide, how long can she make the rug if she wishes to use all the binding on the perimeter of the rug?

16. The radius of a pizza is 8 inches. Find the area of the pizza. (Use 3.14 as an approximation for π.)

17. The circumference of a circular garden is 628 feet. Find the area of the garden. (use 3.14 as an approximation for π.)

18. A water tank is a right circular cylinder. The tank has a radius of 6 meters and a volume of 1356.48 cubic meters. Find the height of the tank. (Use 3.14 as an approximation for π.)

19. A tent has the shape of a right pyramid. The volume is 200 cubic feet and the height is 12 feet. Find the area of the floor of the tent.

20. A spherical balloon has a radius of 9 centimeters. Find the amount of air required to fill the balloon. (Use 3.14 as an approximation for π.)

21. Linda invests $5000 at 6% simple interest and earns $450. How long did Linda invest her money?

22. Find the height of an ice cream cone if the diameter is 6 centimeters and the volume is 37.68 cubic centimeters. (Use 3.14 as an approximation for π. Round answer to the nearest hundredth.)

Objective 3 Solve problems about vertical angle and straight angles.

Find the measure of each marked angle.

23.

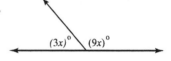

24.

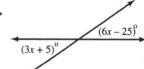

25.

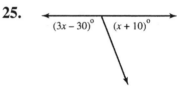

26.

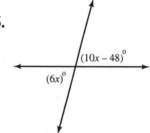

27.

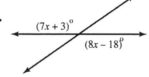

28.

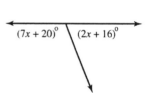

29.

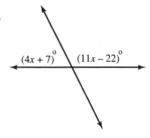

30.

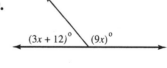

31.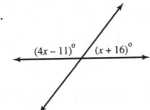

32.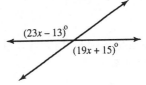

Objective 4 **Solve a formula for a specified variable.**

Solve each formula for the specified variable.

33. $V = LWH$ for H

34. $A = p + prt$ for p

35. $S = \dfrac{a}{1-r}$ for r

36. $A = \frac{1}{2}bh$ for h

37. $S = 2\pi rh + 2\pi r^2$ for h

38. $a_n = a_1 + (n-1)d$ for n

39. $P = A - Art$ for A

40. $V = \pi r^2 h$ for h

41. $A = \frac{1}{2}(b+B)h$ for b

42. $S_n = \frac{n}{2}(a_1 + a_n)$ for a_1

43. $C = \frac{5}{9}(F - 32)$ for F

44. $d = gt^2 + vt$ for v

45. $V = \frac{1}{3}\pi r^2 h$ for h

46. $S = (n-2)180$ for n

2.5 Mixed Exercises

Solve each formula for the specified variable. Then find the value of the specified variable, using the given variables.

47. $P = 2L + 2W$ for L; $P = 262$, $W = 20$

48. $V = LWH$ for H; $V = 80$, $L = 8$, $W = 2$

49. $V = \frac{1}{3}\pi r^2 h$ for h; $V = 78.50$, $r = 5$, $\pi = 3.14$ (Round answer to the nearest hundredth.)

50. $A = \frac{1}{2}(b+B)h$ for b; $A = 30$, $B = 9$, $h = 5$

51. $S = \dfrac{a}{1-r}$ for a; $S = 60$, $r = .4$

52. $P = A - Art$ for r; $P = 1320$, $A = 1500$, $t = 3$

Solve each problem.

53. The radius of a spherical balloon is 4 inches. Find its volume. (Use 3.14 as an approximation for π.)

54. Mr. Fixxall has 1600 meters of fencing material to enclose a rectangular field. The width of the field will be 250 meters. Find the length.

55. The perimeter of a triangle is 46 centimeters. One of the sides is 18 centimeters long, and the other two sides are of equal length. Find the length of each equal side.

56. The perimeter of a rectangular garden is 2800 feet. The width is 430 feet. Find the length.

57. The area of a triangular lot is 490 square meters. If the height is 28 meters, find the base of the lot.

58. A soup can has a height of 6 centimeters and a diameter of 4 centimeters. How much soup will the can hold? (Use 3.14 as an approximation for π.)

59. A pair of vertical angles have measures $(5x-20)^\circ$ and $(3x+14)^\circ$. Find the measures of these two angles.

60. Two angles with measures of $(3x+42)^\circ$ and $(10x-31)^\circ$ form a straight angle. Find the measures of these two angles.

SOLVING EQUATIONS AND INEQUALITIES

2.6   Ratio, Proportion, and Percent

Objective 1   Write ratios.

Write a ratio for each word phrase. Write fractions in lowest terms.

1. 8 men to 3 men

2. 14 marbles to 18 marbles

3. 10 days to 2 weeks

4. 4 gallons to 2 pints

5. 9 dollars to 48 quarters

6. 9 hours to 2 days

7. 10 inches to 2 yards

8. 23 yards to 10 feet

9. 5 months to 2 years

10. 40 hours to 3 weeks

11. 3 ounces to 1 pound

12. 45 cents to $9

13. 40 centimeters to 2 meters

14. 175 years to 2 centuries

Zagara's supermarket was surveyed and the following prices were charged for items in various sizes. Find the best buy (based on price per unit) for each of the following items.

15. Rice:
     1-pound box: $1.29
     2-pound box: $2.31
     3-pound box: $3.32
     5-pound box: $4.44

16. Tomato catsup
     14-ounce size: $ .93
     32-ounce size: $1.92
     44-ounce size: $2.59
     64-ounce size: $3.45

17. Trash bags
     10-count box: $1.25
     15-count box: $1.60
     20-count box: $1.99
     25-count box: $2.49

18. Corn oil
     18-ounce bottle: $1.27
     32-ounce bottle: $2.40
     48-ounce bottle: $3.19
     64-ounce bottle: $4.43

19. Applesauce
     8-ounce jar: $ .59
     16-ounce jar: $ .96
     24-ounce jar: $1.31
     48-ounce jar: $1.99

20. Freeze-dried coffee
     2-ounce size: $2.21
     4-ounce size: $3.78
     7-ounce size: $4.20
     10-ounce size: $6.14

Objective 2 Solve proportions.

Solve each equation.

21. $\dfrac{25}{3} = \dfrac{125}{x}$

22. $\dfrac{4}{r} = \dfrac{30}{12}$

23. $\dfrac{x}{18} = \dfrac{2}{3}$

24. $\dfrac{z}{20} = \dfrac{25}{125}$

25. $\dfrac{5}{m} = \dfrac{12}{5}$

26. $\dfrac{m}{5} = \dfrac{m-2}{2}$

27. $\dfrac{w+4}{6} = \dfrac{w+10}{8}$

28. $\dfrac{x+8}{x-9} = \dfrac{1}{4}$

29. $\dfrac{6y-4}{y} = \dfrac{11}{5}$

30. $\dfrac{4}{z+1} = \dfrac{2}{z+7}$

31. $\dfrac{3x+4}{x-2} = \dfrac{1}{3}$

32. $\dfrac{s+2}{s+8} = \dfrac{9}{14}$

Objective 3 Solve applied problems using proportions.

Solve the following problems involving proportions.

33. Ginny can type 8 pages of her term paper in 30 minutes. How long will it take her to type the paper if it has 20 pages?

34. If 6 typewriter ribbons cost $40.50, how much will 4 ribbons cost?

35. On a road map, 6 inches represents 50 miles. How many inches would represent 125 miles?

36. A certain lawn mower uses 5 tanks of gas to cut 18 acres of lawn. How many acres could be cut using 12 tanks of gas?

37. If 3 ounces of medicine must be mixed with 10 ounces of water, how many ounces of medicine must be mixed with 15 ounces of water?

38. A certain lawn mower uses 7 tanks of gas to cut 15 acres of lawn. How many tanks of gas are needed to cut 30 acres of lawn?

39. If 12 rolls of tape cost $4.60, how much will 15 rolls cost?

40. If four pounds of fertilizer will cover 50 square feet of garden, how many pounds would be needed for 125 square feet?

41. A garden service charges $30 to install 50 square feet of sod. Find the charge to install 225 square feet.

42. The charge to move a load of freight 700 miles is $100. Find the charge to move the freight 1750 miles.

Objective 4   **Find percentages and percents.**

Answer each of the following.

43. Find 45% of 420.

44. Find 150% of 60.

45. What is 2.5% of 3500?

46. What percent of 480 is 34.8?

47. What percent of 75 is 450?

48. What percent of 5200 is 104?

49. 30% of what number is 97.2?

50. 2.75% of what number is 20.625?

51. What is 6.5% of 230?

52. 14.96 is what percent of 88?

Solve each problem.

53. The number of students enrolled in a calculus course is 145. If 40% of these students are female, how many are female?

54. A family of four with a monthly income of $2650 spends 75% of its income and saves the rest. Find the *annual* savings for this family.

55. Pablo recently bought a duplex for $144,000. He expects to earn $6120 per year on this investment. What percent of the purchase price will he earn?

56. Ruth calculated that she had traveled 805 miles by bus on a trip that covered 1150 miles. What percent of the trip was by bus?

2.6 Mixed Exercises

Write a ratio for each word phrase. Write fractions in lowest terms.

57. 2 dollars to 7 dimes

58. 90 minutes to 2 days

59. 30 minutes to 4 hours

60. 20 ounces to 7 pounds

61. 7 quarters to 2 dollars

62. 5 pints to 2 quarts

Find the best buy (based on price per unit) for the following items.

63. Chilled orange juice
32-ounce carton: $1.55
96-ounce carton: $2.46
64-ounce carton: $3.89

64. Liquid dishwashing detergent
12-ounce size: $1.09
22-ounce size: $1.79
32-ounce size: $2.39
42-ounce size: $3.07

Solve each equation.

65. $\dfrac{k}{24} = \dfrac{6}{144}$

66. $\dfrac{32}{s} = \dfrac{8}{6}$

67. $\dfrac{m}{18} = \dfrac{8}{63}$

68. $\dfrac{t-2}{t+4} = \dfrac{1}{5}$

69. $\dfrac{q+2}{q+4} = \dfrac{7}{12}$

70. $\dfrac{p}{32} = \dfrac{p+2}{48}$

Solve each problem.

71. A pair of jeans with a regular price of $45 is on sale this week at 45% off. Find the amount of discount and the sale price of the jeans.

72. An advertisement for a DVD player gives a sale price of $175.50. The regular price is $225. Find the percent discount on this DVD player.

73. A garden service charges $45 to install 50 square feet of sod. Find the charge to install 125 square feet.

74. If 12 heads of lettuce cost $11.25, how much will 16 heads cost?

SOLVING LINEAR EQUATIONS AND INEQUALITIES

2.7 Solving Linear Inequalities

Objective 1 Graph inequalities on a number line.

Graph each inequality on a number line.

1. $x \geq 3$ 2. $7 < a$ 3. $n \leq 7$ 4. $y \geq -2$

5. $-5 \leq r$ 6. $x \geq 0$ 7. $-4 \leq x < 4$ 8. $-3 \leq y < 0$

9. $-1 < x < 3$ 10. $-6 < m < -2$ 11. $8 \leq k \leq 10$ 12. $-3 < a \leq 2$

Objective 2 Solve linear inequalities using the addition property.

Solve each inequality and graph the solutions.

13. $j + 6 \leq 11$ 14. $4m - 4 < 5m$ 15. $-2 + 8b \geq 7b - 1$

16. $t + 11 > 14$ 17. $y - 7 > -12$ 18. $5a + 3 \leq 6a$

19. $14 + 23a \geq 24a + 18$ 20. $6 + 3x < 4x + 4$ 21. $3 + 5p \leq 4p + 3$

22. $9 + 8b > 9b + 11$

Objective 3 Solve linear inequalities using the multiplication property.

Solve each inequality and graph the solutions.

23. $2x \leq 10$ 24. $-2s > 4$ 25. $-7q \geq 35$

26. $\frac{1}{2}r > 5$ 27. $5r > -20$ 28. $4k \geq -16$

29. $-6k \leq 0$ 30. $\frac{3}{5}n \geq 0$ 31. $-\frac{2}{3}z > 4$

32. $-.04t \leq .2$ 33. $-5t \leq -35$ 34. $-9m > -36$

Objective 4 Solve linear inequalities with three parts.

Solve each inequality.

35. $-2 < y - 3 < 6$ 36. $-6 < k + 2 < 8$

37. $10 < z + 5 < 14$　　　　　　**38.** $-4 \leq a + 5 < -2$

39. $-4 < 6 - 2x < -2$　　　　　　**40.** $4 > 3a + 4 > -4$

Solve each inequality and graph the solutions.

41. $7m - 8 \geq 5m$　　　　　　　**42.** $p - 5 - 4p > 8 - p$

43. $5p - 5 - p > 7p - 2$　　　　**44.** $5(x+4) + 2x < 2(3x-1) + 8$

45. $5(y+3) - 5y > 3(y+1) + 4$　　**46.** $3 - \frac{1}{4}z \leq 2 + \frac{3}{8}z$

47. $4 - \frac{1}{3}y \leq 6 + \frac{2}{3}y$　　　　**48.** $3(z+1) \leq 5(2z-4) + 2$

Objective 5　 **Solve applied problems using linear inequalities.**

Use an inequality to solve each problem.

49. Faustino sold two antique desks for $280 and $305. How much should he charge for the third in order to average at least $300 per desk?

50. Find every number such that one third the sum of that number and 24 is less than or equal to 10.

51. Lauren has grades of 98 and 86 on her first two chemistry quizzes. What must she score on her third quiz to have an average of at least 91 on the three quizzes?

52. David has one more paper to write this term in his composition class. If he has scored 92, 84, 90, and 78 on papers so far, what grade must he receive on the last paper to have an average of at least 85 for the five papers?

53. Nina has a budget of $230 for gifts for this year. So far she has bought gifts costing $47.52, $38.98, and $26.98. If she has three more gifts to buy, find the average amount she can spend on each gift and still stay within her budget.

54. Linda earns $6 an hour on one part-time job and $5.75 on another. How much must she earn on her third part-time job for her average hourly pay to be at least $6.50?

55. Ruth tutors mathematics in the evenings in an office for which she pays $600 per month rent. If rent is her only expense and she charges each student $40 per month, how many students must she teach to make a profit of at least $1600 per month?

56. Faustino is paid 32 cents per newspaper to deliver papers for the *Phila. Inquirer*. How many papers must be deliver in an hour if he is to make at least $8 per hour?

57. Two sides of a triangle are equal in length, with the third side 8 feet longer than one of the equal sides. The perimeter of the triangle cannot be more than 38 feet. Find the largest possible value for the length of the equal sides.

58. The perimeter of a triangle must be no more than 45 centimeters. One side of the triangle is 16 centimeters and a second side is 12 centimeters. Find the largest possible length for the third side.

2.7 Mixed Exercises

Solve each inequality and graph the solutions.

59. $y + 2 \geq 9$

60. $7k \geq -28$

61. $4x \leq 3x + 3$

62. $-5k \leq 0$

63. $-1 + 12b \leq 11b - 4$

64. $\frac{3}{5}m + 2 > \frac{1}{3}m$

65. $10p + 4 > 6p - 8$

66. $7w - 6 \geq 12w - 15$

67. $.8x + 3 \leq .4x + 4$

68. $20 - 5x > 17 - 2x$

69. $4(z - 3) + 3(2z + 5) < -7$

70. $6(x - 3) + 4 \leq 3(x - 2) + 2x$

71. $4(y - 6) + 18 \leq 3(y + 2) - 12$

72. $-9t + 4(t - 2) > t - (6 + 5t) + 9$

73. $\frac{3}{4}(r - 5) > \frac{1}{6}(r + 2)$

74. $\frac{1}{2}(x - 5) \leq \frac{3}{10}(x - 3)$

Use inequalities to solve the following problems.

75. If 7 is subtracted from three times a number, the result is less than 8. Find all numbers that satisfy this condition.

76. Carla sold 10 cars in April and 29 cars in May. How many cars must she sell in June in order to average at least 16 cars per month?

77. If twice the sum of a number and 7 is subtracted from three times the number, the result is more than –9. Find all such numbers.

78. The perimeter of a triangle must be no less than 129 centimeters. One side of the triangle is 33 centimeters, and a second side is 49 centimeters. Find the smallest possible length for the third side.

79. Nina earned $450 in January, $320 in February, and $665 in March as a tutor. How much must she earn in April in order to average at least $500 per month for the four months from January through April?

80. The Thomas Jefferson P.T.O. wishes to build a playground with a rectangular fence. If 136 meters of fencing are available, and the width of the playground must be 15 meters, find all possible lengths of the playground.

Chapter 3

GRAPHS OF LINEAR EQUATIONS AND INEQUALITIES IN TWO VARIABLES

3.1   Reading Graphs; Linear Equations in Two Variables

Objective 1   Interpret graphs.

The pie chart below shows how the Patel family used its income in 2000. The total family income was $58,000. Use this chart to answer the questions in Exercises 1 – 3.

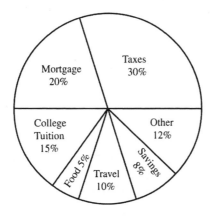

1.   What percentage of the family's income was spent on the combination of the following expenses: taxes, food, and travel?

2.   How much money was spent on mortgage payments?

3.   How much more money was spent on taxes than went to savings?

Use the bar graph to answer the questions in Exercise 4 – 6.

4. In which year were auto retail sales the lowest?

5. Between what two years did auto retail sales decrease?

6. Between what two years did auto retail sales increase by the largest amount?

The line graph below shows the rate of inflation during recent years in the United States. Use this graph to answer the questions in Exercises 7 – 9.

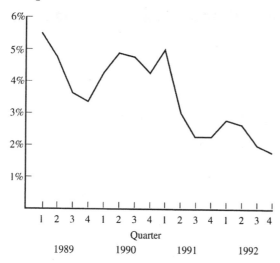

7. What is the highest inflation rate shown on the graph? When did it occur?

8. What is the lowest inflation rate shown on the graph? When did it occur?

9. During which quarter did the greatest decrease in inflation rate occur?

The graphs below show the total number of degrees awarded by Jefferson University for the years 1990 – 1995 and the distribution of degrees awarded over this period. Use these graphs to answer the questions in Exercises 10 – 15.

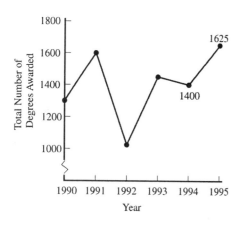 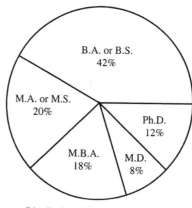

Distribution of Degrees, 1990 – 1995

10. Between which two years did the total number of degrees awarded show the greatest decline?

11. About how many students received master's degrees (M.A. or M.S.) in 1994?

12. About how many more students received M.B.A. degrees in 1995 than 1994?

13. Between which two years did the total number of degrees awarded show the smallest change?

14. About how many students received doctoral degrees (M.D., or Ph.D.) in 1994?

15. About how many more students received Ph.D. degrees in 1995 than in 1994?

The graphs below show the usage of a mathematics help center by subject and by day of the week. Use these graphs to answer the questions in Exercises 16 – 21.

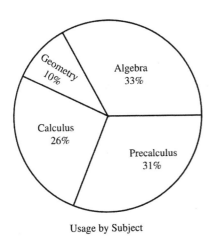

Usage by Subject

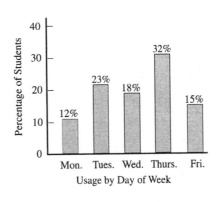

16. On which day was the center used one-and-a-half times as much as Monday?

17. If 340 students came for geometry help, how many students came for calculus help?

18. If 450 students came to the center each week, how many students used the center on Wednesday?

19. If 4500 students used the math center during the school year, how many of these students were not enrolled in geometry or pre-calculus courses? Assume each student is enrolled in only one math class.

20. Which day had the greatest decrease in usage of the center as compared to the previous day?

21. If 360 students used the center on Mondays during the fall semester, how many used the center on Fridays?

The line graph below shows industrial production of three countries. Use this graph to answer the questions in Exercises 22 – 24.

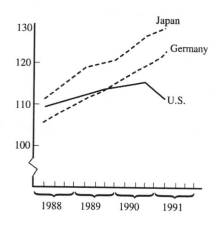

22. Which country had the smallest change in industrial production over the entire period?

23. Which country showed the greatest increase over the entire period?

24. Which country showed the most steady rate of growth in industrial production over the entire period?

Objective 2   Write a solution as an ordered pair.

Write each solution as an ordered pair.

25. $x = 4$ and $y = 7$ **26.** $x = -7$ and $y = 2$

27. $x = -4$ and $y = 0$ **28.** $x = 2$ and $y = 7$

29. $x = \frac{1}{3}$ and $y = -9$ **30.** $x = -2$ and $y = -3$

31. $x = 4$ and $y = 7$ **32.** $y = \frac{1}{3}$ and $x = 0$

33. $x = .2$ and $y = .3$ **34.** $x = 9$ and $y = -7$

Objective 3   Decide whether a given ordered pair is a solution of the given equation.

Decide whether the given ordered pair is a solution of the given equation.

35. $3x + 2y = 4;\ (0, 2)$ **36.** $5x - 2y = 6;\ (2, -2)$

37. $4x - 3y = 10;\ (1, 2)$ **38.** $2x - 3y = 1;\ \left(0, \frac{1}{3}\right)$

39. $2x = -4y;\ (0, 0)$ **40.** $2x = 3y;\ (3, 2)$

41. $y+3=0;\ (0,-3)$

42. $x=-7;\ (-7,9)$

43. $x-5=1;\ (6,4)$

44. $x=1-2y;\ \left(0,-\tfrac{1}{2}\right)$

Objective 4   Complete ordered pairs for a given equation.

For each of the given equations, complete the ordered pairs beneath it.

45. $y=2x-5$
   (a) $(2,\ \ )$
   (b) $(0,\ \ )$
   (c) $(\ \ ,3)$
   (d) $(\ \ ,-7)$
   (e) $(\ \ ,9)$

46. $y=-2x+4$
   (a) $(1,\ \ )$
   (b) $(-4,\ \ )$
   (c) $(\ \ ,6)$
   (d) $(\ \ ,16)$
   (e) $(\ \ ,-6)$

47. $y=3+2x$
   (a) $(-4,\ \ )$
   (b) $(2,\ \ )$
   (c) $(\ \ ,0)$
   (d) $(-2,\ \ )$
   (e) $(\ \ ,-7)$

48. $5x+4y=10$
   (a) $(2,\ \ )$
   (b) $(4,\ \ )$
   (c) $(\ \ ,3)$
   (d) $(0,\ \ )$
   (e) $(\ \ ,0)$

49. $x=-2$
   (a) $(\ \ ,-2)$
   (b) $(\ \ ,0)$
   (c) $(\ \ ,19)$
   (d) $(\ \ ,3)$
   (e) $\left(\ \ ,-\tfrac{2}{3}\right)$

50. $y=4$
   (a) $(2,\ \ )$
   (b) $(0,\ \ )$
   (c) $(4,\ \ )$
   (d) $(-4,\ \ )$
   (e) $(.75,\ \ )$

Objective 5   Complete the table of values.

Complete the table of ordered pairs for each equation.

51. $4x + y = 6$

x	2		1
y		4	

52. $3x = 2y$

x	2		
y		-3	0

53. $3x + 2y = 4$

x	0		4
y		0	

54. $y = 2x - 4$

x	0	3	
y			8

55. $3x + y = 9$

x	0		-1
y		0	

56. $3x + 2y = 12$

x	0		-8
y		0	

57. $x = -2$

x			
y	0	4	-5

58. $y = 2$

x	-5	0	7
y			

59. $x - 4 = 0$

x			
y	-8	4	8

60. $y - 4 = 0$

x	-6	0	6
y			

61. $3x - 4y = -6$

x	y
0	
	0
2	

62. $4x + 3y = 12$

x	y
0	
	0
	-1

63. $-7x + 2y = -14$

x	y
	0
0	
3	

64. $3x + 4y = 12$

x	y
	0
0	
	-3

65. $y - 4 = 0$

x	y
-4	
0	
6	

66. $2x + 5 = 7$

x	y
	-3
	0
	5

Objective 6   Plot ordered pairs.

Plot the following ordered pairs on a coordinate system.

67. $(7, 1)$ **68.** $(-2, 4)$ **69.** $(5, 6)$ **70.** $(-2, -7)$

71. $(4,-2)$ **72.** $(-3,4)$ **73.** $(1,0)$ **74.** $(0,-4)$

75. $(-5,0)$ **76.** $(0,2)$ **77.** $(0,0)$ **78.** $(2,5)$

Without plotting the given point, name the quadrant in which it lies.

79. $(1,-2)$ **80.** $(5,-4)$ **81.** $(-7,-2)$ **82.** $(3,-2)$

83. $\left(\frac{1}{2},\frac{4}{5}\right)$ **84.** $(0,-2)$ **85.** $(3,0)$ **86.** $(0,0)$

3.1 Mixed Exercises

The graphs below show the population of Riverville for the years 1975 – 1995 and the average age distribution of the population over this period. Use these graphs to answer the questions in Exercises 87 – 92.

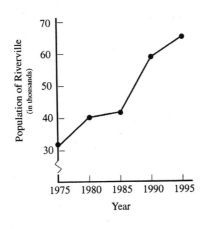 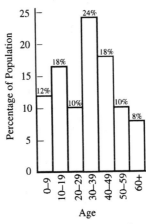

87. During which 5-year period did the population show the greatest change?

88. About how many people were in the 40–49 age group in 1990?

89. In 1985, about how many more people were in the 40–49 age group than in the   20–29 age group?

90. During which five-year period did the population show the smallest change?

91. About how many people were 40 years or older in 1990?

92. In 1985, about how many more people were in the 30–39 age group than in the   10–19 age group?

Decide whether the given ordered pair is a solution of the given equation.

93. $3x+2y=8$; $(4,0)$ **94.** $3x+2y=5$; $(1,5)$ **95.** $y=2x$; $(3,6)$

96. $3x+y=4$; $(2,0)$ **97.** $3x=2y$; $(4,6)$ **98.** $x+4=y$; $(0,-4)$

Complete the given ordered pairs for each equation.

99. $2y = -4x + 4$ $\quad$ $(6, \ )$ $\left(-\frac{1}{2}, \ \right)$ $(\ , 0)$

100. $3x + 4y = 12$ $\quad$ $(0, \ )$ $(\ , 0)$ $(2, \ )$

101. $2x + y = 4$ $\quad$ $(3, \ )$ $(\ , 5)$ $(2.5, \ )$

102. $y + 3 = -1$ $\quad$ $(0, \ )$ $(-1, \ )$ $(100, \ )$

Complete the table of ordered pairs for each equation.

103. $-2x + y = 4$

x	3		−4
y		0	

104. $4x + 3y = 5$

x	0		1
y		0	

105. $x + 3 = -4$

x			
y	−5	0	10

106. $5x + 2y = 10$

x		0	4
y	0		

107. $5x + 6y = 30$

x	y
0	
	0
3	

108. $-2x + 4y = 20$

x	y
	0
0	
	3

GRAPHS OF LINEAR EQUATIONS AND INEQUALITIES IN TWO VARIABLES

3.2 Graphing Linear Equations in Two Variables

Objective 1 Graph linear equations by plotting ordered pairs.

Complete the ordered pairs for each equation. Then graph the equation by plotting the points and drawing a line through them.

1. $x + y = 3$
    (0,)
    (,0)
    (2,)

2. $y = x + 7$
    (0,)
    (,0)
    (−4,)

3. $y + 3 = 0$
    (0,)
    (4,)
    (−3,)

4. $x + y = -2$
    (0,)
    (,0)
    (3,)

5. $2y - 4 = x$
    (0,)
    (,0)
    (−2,)

6. $x - 4 = 0$
    (,0)
    (,−2)
    (,3)

7. $x - y = 4$
    (0,)
    (,0)
    (−2,)

8. $y = 3x - 2$
    (0,)
    (,0)
    (2,)

9. $x = -2y + 2$
    (0,)
    (,0)
    (−2,)

10. $x - y = -1$
    (0,)
    (,0)
    (4,)

11. $y = 2x - 3$
    (0,)
    (,0)
    (1,)

12. $2x + 3y = 6$
    (0,)
    (,0)
    (−3,)

Objective 2 Find intercepts.

Find the intercepts for the graph of each equation.

13. $-5x + 2y = 10$

14. $3x + 2y = 12$

15. $2x + 4y = 0$

16. $4x + 5y = 8$

17. $5x - 2y = 10$

18. $4x + 3y = 9$

19. $3x + 2y = -2$ **20.** $5x - 3y = 12$ **21.** $2x + 9y = -9$

22. $3x + 4y = 9$

Find the intercepts for each equation. Then graph the equation.

23. $3x + y = 6$ **24.** $6x + 5y = 15$ **25.** $x + 2y = -3$

26. $5x + 6y = -30$ **27.** $4x - y = 4$ **28.** $2x - 3y = 6$

29. $3x + 4y = -12$ **30.** $x + 2y = 4$ **31.** $2x - 3y = 6$

32. $5x - 2y = -10$

| Objective 3 | **Graph linear equations where the intercepts coincide.**

Graph each equation.

33. $3x - y = 0$ **34.** $2x + y = 0$ **35.** $3x + 4y = 0$

36. $x + 5y = 0$ **37.** $3x - 2y = 0$ **38.** $-4x + 5y = 0$

39. $x + y = 0$ **40.** $y = 2x$ **41.** $4x = 3y$

42. $5x + 2y = 0$

| Objective 4 | **Graph linear equations of the form** $y = k$ **or** $x = k$.

Graph each equation.

43. $x = 3$ **44.** $x = -3$ **45.** $x + 4 = 0$ **46.** $y = 0$

47. $y = 2$ **48.** $x - 1 = 0$ **49.** $y = -2$ **50.** $y + 3 = 0$

51. $y - 4 = 0$ **52.** $x = 0$

| Objective 5 | **Use a linear equation to model data.**

Solve each problem.

53. The enrollment at Lincolnwood High School decreased during the years 1990 to 1995. If $x = 0$ represents 1990, $x = 1$ represents 1991, and so on, the number of students enrolled in the school can be approximated by the equation

$$y = -85x + 2435.$$

Use this equation to approximate the number of students in each year from 1990 through 1995.

54. The profit y in millions of dollars earned by a small computer company can be approximated by the linear equation

$$y = .63x + 4.9,$$

where $x = 0$ corresponds to 1994, $x = 1$ corresponds to 1995, and so on. Use this equation to approximate the profit in each year from 1994 through 1997.

55. The number of band instruments sold by Elmer's Music Shop can be approximated by the equation

$$y = 325 + 42x,$$

where y is the number of instruments sold and x is the time in years, with $x = 0$ representing 1993. Use this equation to approximate the number of instruments sold in each year from 1993 through 1996.

56. Suppose that the demand and price for a certain model of calculator are related by the equation

$$y = 45 - \tfrac{3}{5}x,$$

where y is the price (in dollars) and x is the demand (in thousands of calculators). Assuming that this model is valid for a demand up to 50,000 calculators, find the price at each of the following levels of demand.

(a) 0 calculators

(b) 5000 calculators

(c) 20,000 calculators

(d) 45,000 calculators

3.2 Mixed Exercises

Complete the ordered pairs for each equation. Then graph the equation by plotting the points and drawing a line through them.

57. $2x + 5y = 20$

(0,)

(,0)

(5,)

58. $x = 3y - 6$

(0,)

(,0)

(,4)

59. $x = 4y - 1$

(3,)

(,0)

(,−1)

60. $2x = y - 4$

(0,)

(,0)

(−4,)

61. $x = 2y + 1$

(0,)

(,0)

(,−2)

62. $y - 3 = x$

(0,)

(,0)

(,4)

Find the intercepts for the graph of each equation. Then graph the equation.

63. $3x + 2y = 12$

64. $3x - 2y = 8$

65. $2x + 5y = -10$

66. $x - 6 = 0$

67. $y + 5 = 0$

68. $4x - 7y = -8$

69. $4x - 3y = 0$

70. $2x + 3y = 0$

71. $4x + 3y = 6$

72. $3x - 5y = 0$

GRAPHS OF LINEAR EQUATIONS AND INEQUALITIES IN TWO VARIABLES

3.3 Slope of a Line

Objective 1 Find the slope of a line given two points.

Find the slope of each line.

1. Through $(4,3)$ and $(3,5)$

2. Through $(2,3)$ and $(6,7)$

3. Through $(-3,2)$ and $(7,4)$

4. Through $(5,-2)$ and $(2,7)$

5. Through $(2,-4)$ and $(-3,-1)$

6. Through $(7,2)$ and $(-7,3)$

7. Through $(0,-5)$ and $(7,2)$

8. Through $(-7,-7)$ and $(2,-7)$

9. Through $(-4,-4)$ and $(-2,-2)$

10. Through $(0,0)$ and $(6,-7)$

11. Through $(-4,6)$ and $(-4,-1)$

12. Through $(2,-7)$ and $(-2,1)$

Objective 2 Find the slope from the equation of a line.

Find slope of each line.

13. $y=-5x$

14. $y=\frac{1}{2}x+5$

15. $y=-\frac{2}{5}x-4$

16. $y=-\frac{4}{7}x+9$

17. $4y=3x+7$

18. $3y=2x-1$

19. $2x+7y=7$

20. $7y-4x=11$

21. $4x-3y=0$

22. $y=-4$

23. $x=0$

24. $3x=4y$

Objective 3 **Use slope to determine whether two lines are parallel, perpendicular, or neither.**

In each pair of equations, give the slope of each line, and then determine whether the two lines are parallel, perpendicular, or neither.

25. $y = -5x - 2$
$y = 5x + 11$

26. $y = 4x + 4$
$y = 3 - \frac{1}{4}x$

27. $-x + y = -7$
$x - y = -3$

28. $2x + 2y = 7$
$2x - 2y = 5$

29. $4x + 2y = 8$
$x + 4y = -3$

30. $9x + 3y = 2$
$x - 3y = 5$

31. $4x + 2y = 7$
$5x + 3y = 11$

32. $8x + 2y = 7$
$x = 3 - y$

33. $y + 4 = 0$
$y - 7 = 0$

34. $y = 9$
$x = 0$

35. $4x - 3y = 4$
$4x - 3y = 2$

36. $6x + 5y = 8$
$6x - 5y = 24$

Objective 4 **Solve problems involving average rate of change.**

Solve each problem.

37. Suppose the sales of a company are given by the linear equation $y = 1250x + 10,000$, where x is the number of years after 1980, and y is the sales in dollars. What is the average rate of change in sales per year?

38. Suppose a man's salary was $15,750 in 1979 and has risen an average of $1500 per year. If the yearly salaries were plotted on a graph, what would be the slope of the line on which they approximately lie?

39. A small company had the following sales during their first three years of operation.

Year	Sales
2003	$82,250
2004	$89,790
2005	$96,100

  (a) What was the rate of change from 2003–2004?

  (b) What was the rate of change from 2004–2005?

  (c) What was the average rate of change from 2003–2005?

40. A plane had an altitude of 8500 feet at 4:02 P.M. and 12,700 feet at 4:39 P.M. What was the average rate of change in the altitude in feet per minute?

41. A ramp is 10 feet high on the high end and 3 feet high on the low end. It covers a horizontal length of 29 feet. What is the average rate of change of the incline?

42. Enrollment in a college was 11,500 two years ago, 10,975 last year, and 10,800 this year.

    (a) What is the average rate of change in enrollment per year for this 3–year period?

    (b) Explain

3.3 Mixed Exercises

Find the slope of each line.

43. Through $(3,7)$ and $(4,9)$

44. Through $(4,9)$ and $(-4,1)$

45. Through $(3,-1)$ and $(2,1)$

46. Through $(-1,5)$ and $(1,5)$

47. Through $(0,2)$ and $(0,-2)$

48. Through $(5,-2)$ and $(-5,2)$

49. $y = \frac{4}{3}x - 2$

50. $y = \frac{2}{3}x - 17$

51. $6x - 2y = 5$

52. $x = -4$

53. $x = 7y$

54. $2x - y = 3$

In each pair of equations, give the slope of each line, and then determine whether the two lines are **parallel, perpendicular,** *or* **neither.**

55. $4x + 4y = 1$
$5x - 4y = 2$

56. $3x + 2y = 1$
$-2x + 4y = 1$

57. $x = 19$
$y = 2$

58. $y = 6$
$y + 2 = 9$

59. $8x - 9y = 8$
$-8x + 9y = 2$

60. $3x - 11y = 4$
$3x + 11y = 2$

Solve each problem.

61. A company had 41 employees during the first year of operation. During their eighth year, the company had 79 employees. What was the average rate of change in the number of employees per year?

62. A state had a population of 3,105,900 in 1985. The population is declining at an average rate of 5200 people a year. At that rate, predict the population for the year 1999.

GRAPHS OF LINEAR EQUATIONS AND INEQUALITIES IN TWO VARIABLES

3.4 Equations of Lines

Objective 1 Write the equation of a line given its slope and y-intercept.

Write an equation in standard form for each line.

	Slope	y-intercept			Slope	y-intercept
1.	2	$(0, -5)$		**7.**	$\dfrac{3}{5}$	$\left(0, \dfrac{2}{5}\right)$
2.	6	$(0, -2)$				
3.	-4	$(0, 3)$		**8.**	$\dfrac{6}{5}$	$\left(0, -\dfrac{1}{5}\right)$
4.	-5	$(0, 3)$				
5.	$-\dfrac{2}{3}$	$(0, 2)$		**9.**	$\dfrac{7}{3}$	$(0, 9)$
6.	$-\dfrac{1}{4}$	$(0, -3)$		**10.**	0	$(0, 3)$

Objective 2 Graph a line given its slope and a point on the line.

Objective 3 Write the equation of a line given its slope and a point on the line.

Write an equation for each line. Write answers in the form $Ax + By = C$.

	Slope	Point			Slope	Point
				18.	$-\dfrac{2}{3}$	$(1, -5)$
11.	2	$(-4, 1)$				
12.	4	$(2, 6)$		**19.**	$-\dfrac{3}{4}$	$(-1, -3)$
13.	5	$(3, -6)$				
14.	1	$(-4, 5)$		**20.**	$-\dfrac{4}{5}$	$(3, -2)$
15.	-5	$(-4, 19)$				
				21.	undefined	$(3, 0)$
16.	-3	$(2, -3)$		**22.**	0	$(-5, 2)$
17.	$-\dfrac{1}{2}$	$(-3, 2)$		**23.**	0	$(3, -4)$
				24.	undefined	$(0, 6)$

25. undefined   (2, 7)

26.   0   (3, 0)

Objective 4 Write the equation of a line given two points on the line.

Write an equation in standard form of the line passing through the given pair of points.

27. (4, 9), (3, 8)

28. (7, 1), (6, 5)

29. (3, 7), (5, 4)

30. (2, –1), (5, –2)

31. (–6, 2), (–4, 1)

32. (3, –2), (–1, 5)

33. (–3, –2), (–5, –1)

34. (–1, –4), (–2, –3)

35. (9, 1), (–9, 1)

36. (3, –5), (–4, –5)

37. (0, 2), (0, –6)

38. (–1, –7), (–1, 8)

Objective 5 Write the equation of a line parallel or perpendicular to a given line.

Write an equation in standard form for each line.

39. parallel to $x - y = 4$, through (4, –7)

40. parallel to $2x + 3y = -12$, through (9, –3)

41. parallel to $2x + 6y = 5$, through (1, –2)

42. parallel to $4x - 3y = 8$, through (–2, 3)

43. parallel to $5x + y = 6$, through (0, 4)

44. perpendicular to $x - 3y = 0$, through (–10, 2)

45. perpendicular to $5x + y = 8$, through (2, –1)

46. perpendicular to $3x - 2y = 6$, through (5, –3)

47. perpendicular to $x = 4$, through (–1, 7)

48. perpendicular to $y = -1$, through (2, 5)

49. parallel to $y = 2$, through (–4, 6)

50. parallel to $x + 1 = 3$, through (–3, 5)

Objective 6 Apply concepts of linear equations to real data.

For each situation,
  (a) *Write an equation in the form $y = ax$;*
  (b) *Give the three ordered pairs associated with the equation for x-values 0, 5, and 10.*

51. x represents the number of minutes for a long distance call at $.13 per minute, and y represents the total cost of the call (in dollars).

52. x represents the number of rows of chairs, with 8 chairs in each row, set up for a concert, and y represents the total available seating.

53. x represents the number of lines of type in an ad at $1.25 per line, and y represents the total charge (in dollars).

For each situation,
  (a) *Write an equation in the form $y = ax + b$;*
  (b) *Give the three ordered pairs associated with the equation for x-values 0, 5, and 10.*

54. A long distance phone call costs $.35 plus $.13 per minute for each minute of the call. Let x represent the number of minutes so that y represents the total cost of the call (in dollars).

55. A music teacher set up rows of chairs for a concert. There were 8 chairs in each row, plus 15 special reserved seats up front for faculty. Let x represent the number of rows of chairs so that y represents the total number of guests who can be seated.

56. To run a newspaper ad, there is a $25 set up fee plus a charge of $1.25 per line of type in the ad. Let x represent the number of lines of type in the ad so that y represents the total cost of the ad (in dollars).

Write a linear equation and solve it in order to solve the problem.

57. Refer to Exercise 64. Suppose the call cost $1.91. How long was the call (in minutes)?

58. Refer to Exercise 65. Suppose 207 people were seated for the concert. How many rows of seats were there?

59. Refer to Exercise 66. A newspaper ad cost $62.50. How many lines long was the ad?

3.4   Mixed Exercises

Write an equation in standard form of each line satisfying the given conditions.

60. slope, $-\dfrac{5}{8}$; y-intercept, $\left(0, -\dfrac{2}{3}\right)$

61. through (2, 5) and (3, 6)

62. slope, $-\dfrac{4}{7}$; through (0, –8)

63. undefined slope; through (–3, 5)

64. parallel to $3x + 4y = 4$; through (–8, 4)

65. slope, 0; y-intercept, (0, –5)

66. slope, 0; through (0, 2)

67. through (–2, –4) and (–2, –7)

68. perpendicular to $x + y = 4$; through (2, 5)

69. through (9, –2) and (10, –5)

70. slope, –3; through (–4, 11)

71. through (6, 5) and (–6, 5)

Find the slope and y-intercept of each line.

72. $2x + 7y = 14$

73. $3x - 2y = 9$

74. $x = -10$

75. $y = 2$

GRAPHS OF LINEAR EQUATIONS AND INEQUALITIES IN TWO VARIABLES

3.5 Graphing Linear Inequalities in Two Variables

Objective 1   Graph linear inequalities in two variables.

Graph each linear inequality.

1. $y \geq x - 1$

2. $x + y \geq 2$

3. $3x - 2y \leq 6$

4. $y \geq -1$

5. $x - 4 \leq -1$

6. $y \leq -\frac{2}{5}x + 2$

7. $y \geq 3x$

8. $x - y \leq -3$

9. $2x + 5y \leq -8$

10. $y \leq x + 4$

Graph each linear inequality.

11. $x + 3y < 3$

12. $3x - 5y > -15$

13. $2x + 5y > -10$

14. $y < x - 3$

15. $y > -x + 2$

16. $x < 2y + 4$

17. $5x + 4y > 20$

18. $2x - 3y < 6$

19. $5x - 2y + 10 < 0$

20. $2 - 3y > x$

Objective 2   Graph a linear inequality with boundary through the origin.

Graph each linear inequality.

21. $y \geq 3x$

22. $y \leq \frac{2}{5}x$

23. $y \geq \frac{1}{3}x$

24. $y \geq x$

25. $3x - 4y \geq 0$

26. $x \geq -4y$

27. $x < 2y$

28. $x > -2y$

29. $x > 4y$

30. $3x - 2y < 0$

3.5 Mixed Exercises

Graph each linear inequality.

31. $y \geq -x + 4$

32. $3x + 2y \leq -6$

33. $x - 4 < 0$

34. $y \leq -\frac{1}{2}x + 6$

35. $y + 4 \geq 0$

36. $y < 3x - 2$

37. $x \geq -5$

38. $y < 3$

39. $y \geq -\frac{1}{2}x$

40. $x < \frac{1}{3}y$

41. $-3x + 5y > 15$

42. $3x - 4y - 12 > 0$

GRAPHS OF LINEAR EQUATIONS AND INEQUALITIES IN TWO VARIABLES

3.6 Introduction to Functions

Objective 1 Define and identify relations and functions.

Decide whether each relation is a function.

1. $\{(1, 3), (1, 4), (2, -1), (3, 7)\}$

2. $\{(-1, 2), (0, 5), (1, 8)\}$

3. $\{(2, -2), (3, -3), (4, -4)\}$

4. $\{(6, -3), (4, -2), (2, -1), (0, 0)\}$

5. $\{(0, 4), (3, 2), (0, 0), (3, 5)\}$

6. $\{(-4, -1), (-3, -2), (-1, 0), (0, 5)\}$

7. $\{(1, 1), (1, 2), (1, 7), (2, 1)\}$

8. $\{(3, 4), (5, 2), (4, 3), (5, 3)\}$

9. $\{(1, 5), (2, 5), (4, 5)\}$

10. $y + x = 9$

11. $x^2 + y^2 = 1$

12. $y \geq 8x - 3$

13. $xy = 7$

14. $x = \sqrt{y+1}$

15. $x = y$

Objective 2 Find domain and range.

Decide whether each relation is a function and give the domain and range of the relation.

16. $\{(-1, 1), (-2, 2), (0, 0)\}$

17. $\{(3, 0), (2, 4), (1, 6), (-1, 3)\}$

18. $\{(-2, -2), (-1, -1), (0, 0), (1, -1)\}$

19. $\{(3, 5), (2, 3), (1, 0)\}$

20. $\{(1, 3), (2, -1), (-1, 4), (1, 4)\}$

21. $\{(2, -4), (1, -2), (-1, 2), (0, 3)\}$

22. $\{(5, 2), (3, -1), (1, -3), (-1, -5)\}$

23. $\{(4, 2), (3, 2), (2, 2), (1, 2), (0, 2)\}$

Decide whether each relation defines y as a function of x. Give the domain.

24. $y = 2x + 5$

25. $y \leq 4x$

26. $y = \sqrt{x-4}$

27. $y^2 = x + 1$

28. $y = \dfrac{1}{x}$

29. $y = 2x^2 + 3$

Objective 3 Identify functions defined by graphs and equations.

Decide which are graphs of functions. Use the vertical line test.

30.

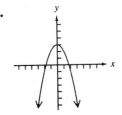

31.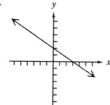

32.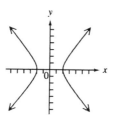

33.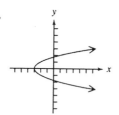

34.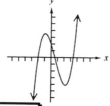

35.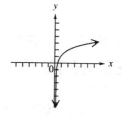

36.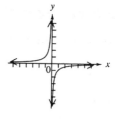

37.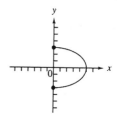

Objective 4   Use function notation.

Find f (–2), f (4), and f (–x).

38. $f(x) = 2x + 5$

39. $f(x) = 6 - 2x$

40. $f(x) = 3x^2$

41. $f(x) = x^2 - 2x$

42. $f(x) = \dfrac{4}{x^2 + 1}$

43. $f(x) = \dfrac{2x + 1}{5}$

Objective 5   Identify linear functions.

Write each equation in the form f (x) = mx + b.

44. $y = x + 2$

45. $2x - y = -2$

46. $x - y = 1$

47. $y + \dfrac{1}{2}x = -2$

48. $x + 4y = 2$

49. $\dfrac{1}{3}y = x$

51. $\dfrac{1}{2}x + \dfrac{1}{3}y = -1$

50. $x + y = 1$

3.6   Mixed Exercises

Decide whether each relation is a function, and give the domain and range of the relation.

52. $\{(0, 1), (1, 3), (2, -4), (4, -8)\}$

53. $\{(-2, -5), (-3, -2), (1, -2)\}$

54. $\{(-4, 1), (-5, 2), (-4, 3), (-3, 4), (-2, 5)\}$

55. $\{(1, 10), (2, 9), (3, 8), (4, 7), (5, -4)\}$

Decide whether the given equation defines y as a function of x. Give the domain.

56. $y = x - 6$

58. $y = |x|$

57. $y = \sqrt{x+2}$

59. $y^2 - 4 = x$

Write each equation in the form f (x) = mx + b. Then find f (-1), f (6), and f (x - 1).

60. $3y + x = 9$

61. $x = \dfrac{1}{2}y + 6$

Chapter 4

SYSTEMS OF EQUATIONS AND INEQUALITIES

4.1 Solving Systems of Linear Equations by Graphing

Objective 1   Decide whether a given ordered pair is a solution of a system.

Decide whether the given ordered pair is a solution of the given system.

1. $(4,1)$
$2x+3y=11$
$3x-2y=9$

2. $(2,-4)$
$2x+3y=6$
$3x-2y=14$

3. $(-3,-1)$
$5x-3y=-12$
$2x+3y=-9$

4. $(4,0)$
$4x+3y=16$
$x-4y=-4$

5. $(-5,-4)$
$x-y=-1$
$4x+y=-24$

6. $(3,-7)$
$5x+y=8$
$2x-3y=26$

7. $(-1,-7)$
$x-y=-6$
$-2x+3y=-19$

8. $(1,-4)$
$-3x+y=-7$
$4x-3y=16$

9. $(-4,-1)$
$5x-2y=6$
$y=-3x-11$

10. $(-1,5)$
$3x+2y=7$
$y=-2x+3$

11. $(4,2)$
$9x-2y=32$
$8x-y=30$

12. $(6,-2)$
$x-y=8$
$2x+3y=6$

13. $(-3,-2)$
$x+8=5$
$y+3=1$

14. $(1,-3)$
$3y=x-10$
$2y=3x-9$

Objective 2   Solve linear systems by graphing.

Solve each system by graphing both equations on the same axes.

15. $x-2y=6$
$2x+y=2$

16. $2x+3y=5$
$3x-y=13$

17. $6x-5y=4$
$2x-5y=8$

18. $3x-y=-7$
$2x+y=-3$

19. $2x=y$
$5x+3y=0$

20. $y-2=0$
$3x+4y=-19$

21. $8x - 5y = -8$
     $2x - y = 0$

22. $x - y = -7$
     $x + 11 = 2y$

23. $3x - 2y = 8$
     $7x + 2y = 12$

24. $2x - 5y = 18$
     $3x = 5y + 22$

25. $x + y = 5$
     $3x - y = -1$

26. $2x - y = -10$
     $3x + 2y = 6$

Objective 3   Solve special systems by graphing.

Solve each system of equations by graphing both equations on the same axes. If the two equations produce parallel lines, write no solution. If the two equations produce the same line, write infinite number of solutions.

27. $8x + 4y = -1$
     $4x + 2y = 3$

28. $x + 2y = 4$
     $8y = -4x + 16$

29. $4x + 3y = 12$
     $6y + 8x = -24$

30. $2x + 3y = 0$
     $6x = -9y$

31. $-3x + 2y = 6$
     $-6x + 4y = 12$

32. $3x + 3y = 8$
     $x = 4 - y$

4.1   Mixed Exercises

Decide whether the given ordered pair is a solution of the given system

33. $(-4, 1)$
     $3x + 2y = -10$
     $4x + y = -15$

34. $(1, -2)$
     $3x - 2y = 7$
     $2x - y = 4$

35. $(5, 3)$
     $3x - 2y = 9$
     $-x + 3y = 4$

36. $(4, -4)$
     $2x - y = 12$
     $3x + 2y = 4$

37. $(2, -3)$
     $3x + y = 9$
     $3x + 4y = -6$

38. $(1, -4)$
     $2x + 5y = -10$
     $-2x + 3y = -14$

Solve each system of equation by graphing both equations on the same axes. If the two equations produce parallel lines, write no solution. If the two equations produce the same line, write infinite number of solutions.

39. $7x - 3y = -21$
     $-21x + 9y = 63$

40. $5x + 3y = 30$
     $10x + 6y = 60$

41. $x + 3y = -2$
     $2x - y = 10$

42. $3x - 2y = 4$
     $7x + y = 15$

43. $2x + 24 = 3y$
     $2y + x = 2$

44. $2x + 3y = 5$
     $x - 2y = -1$

SYSTEMS OF EQUATIONS AND INEQUALITIES

4.2 Solving Systems of Linear Equations by Substitution

Objective 1 Solve linear systems by substitution.

Solve each system by the substitution method. Check each solution.

1. $x+y=7$
    $y=6x$

2. $3x+2y=14$
    $y=x+2$

3. $x+3y=1$
    $x=-5-6y$

4. $y+4x=-1$
    $x=-14-3y$

5. $x+y=9$
    $5x-2y=-4$

6. $x-4y=17$
    $3x-4y=11$

7. $-8x+5y=11$
    $x-y=-1$

8. $3x-21=y$
    $y+2x=-1$

9. $5x+3y=11$
    $x+y=3$

10. $3x-2y=6$
     $x-5y=-11$

11. $3x+4y=2$
     $2x+3y=2$

12. $3x-2y=13$
     $x-5=0$

13. $5x+2y=14$
     $y+2=-1$

14. $x+6y=-1$
     $-2x-9y=0$

15. $6x+8y=10$
     $4y=5-3x$

16. $2x+3y=-4$
     $3x-2y=7$

17. $2x-5y=11$
     $3x-4y=6$

18. $2x+4y=-1$
     $-4x-6y=1$

Solve each system by either the addition method or the substitution method. First simplify equations where necessary. Check each solution.

19. $3x+4y=2x+2y+11$
     $4x-7y=-16$

20. $7x+2y=2x-y+19$
     $4x+3y=2x-2y+0$

21. $2x+8y-4=4y-4x$
     $5x=5-3y$

22. $6x-7y=4x-3y-11$
     $16x+2y=4x+12$

23. $10x+4y=-4x-3y$
     $6x=8y$

24. $10x-y=6x+2y+1$
     $-3x+3y=3x-2y+1$

25. $7x-3y-8=3x-5y$
     $y=2x$

26. $2x+7y=5y-3x+16$
     $2y=-x+y+2$

27. $5y-6x-2y=9y-3x$
     $4y-4x=-2x-8$

28. $5y+5x=x+y+8$
     $5y+3x=2+2y+x+7$

Objective 2 **Solve special systems.**

Objective 3 **Solve linear systems with fractions.**

Solve each system by either the addition method or the substitution method. First clear all fractions. Check each solution.

29. $3x + 5y = 7$
$6x + 10y = 3$

30. $4x + 3y = 2$
$8x + 6y = 4$

31. $\dfrac{x}{6} + \dfrac{y}{6} = 1$
$\dfrac{x}{2} + \dfrac{y}{2} = 2$

32. $\dfrac{5}{3}x + y = 12$
$x + \dfrac{1}{2}y = 7$

33. $\dfrac{5}{4}x - y = -\dfrac{1}{4}$
$-\dfrac{7}{8}x + \dfrac{5}{8}y = 1$

34. $x + \dfrac{5}{3}y = 7$
$\dfrac{5}{6}x + \dfrac{2}{3}y = \dfrac{11}{3}$

35. $x - \dfrac{7}{5}y = \dfrac{6}{5}$
$\dfrac{1}{4}x - \dfrac{1}{2}y = \dfrac{1}{6}$

36. $\dfrac{1}{20}x - \dfrac{1}{15}y = \dfrac{1}{6}$
$\dfrac{1}{6}x + y = 3$

37. $\dfrac{1}{4}x + \dfrac{3}{8}y = -3$
$\dfrac{5}{6}x - \dfrac{3}{7}y = -10$

38. $\dfrac{5x}{4} + \dfrac{2y}{3} = \dfrac{8}{3}$
$\dfrac{2x}{3} - \dfrac{3y}{2} = -6$

39. $\dfrac{x}{2} - \dfrac{y}{3} = -8$
$\dfrac{x}{4} - \dfrac{y}{6} = -4$

40. $\dfrac{x}{3} - \dfrac{y}{2} = -2$
$\dfrac{x}{6} + \dfrac{y}{4} = 0$

41. $\dfrac{9}{2}x - \dfrac{3}{4}y = 3$
$-\dfrac{3}{4}x + \dfrac{1}{8}y = -\dfrac{1}{2}$

42. $\dfrac{8}{3}x - y = \dfrac{7}{5}$
$\dfrac{1}{4}x + \dfrac{3}{8}y = \dfrac{9}{40}$

43. $\dfrac{3}{14}x - \dfrac{1}{7}y = 1$
$-\dfrac{1}{2}x = \dfrac{1}{4}y$

44. $\dfrac{x}{2} - \dfrac{2y}{3} = 4$
$x = \dfrac{4y}{3} + 3$

4.2 Mixed Exercises

Solve each system by the substitution method. Check each solution.

45. $x + 4y = 4$

$y = 6 + x$

46. $4x + 3y = 19$

$x = 2y + 2$

47. $-5x + 3y = 17$

$x - y = -3$

48. $4x + 3y = -10$

$2x + 3y = -2$

49. $4x - 2y = 3$

$-8x + 4y = -6$

50. $3x + 5y = 3$

$6x - 10y = -2$

Solve each system by either the addition method or the substitution method. Check each solution.

51. $7x + 10 - y = 5x - 6y - 6$

$x + 6y - 6 = 7y + 2x - 4$

52. $8x - 2y + 7 = 4x - 5y + 12$

$12 + 3x - 4y = 2x + 2y - 7$

53. $2 + 5x - 7y = -3y - 34$

$7 + 6x - y = 7x + y + 3$

54. $2 + 2x - 5y = 10 - 3y$

$4x + y = 7x - y - 8$

55. $7 + 5x - 7y = -5y - 7$

$7x - y = 9x + y$

56. $4 + 4x - 6y = -4y$

$5x - y = 8x + y + 3$

57. $\dfrac{3}{2}x + \dfrac{2}{3}y = \dfrac{5}{6}$

$\dfrac{2}{3}x - \dfrac{3}{2}y = \dfrac{13}{6}$

58. $\dfrac{5}{3}x + y = \dfrac{1}{3}$

$\dfrac{5}{6}x - \dfrac{9}{2}y = \dfrac{7}{2}$

59. $\dfrac{x}{3} - \dfrac{4y}{9} = \dfrac{7}{9}$

$-\dfrac{3x}{4} + y = -\dfrac{7}{4}$

60. $\dfrac{2x}{3} + \dfrac{5y}{6} = 1$

$\dfrac{4x}{3} + \dfrac{3y}{2} = 3$

61. $\dfrac{3x}{4} + \dfrac{2y}{3} = 5$

$\dfrac{5x}{2} + \dfrac{5y}{3} = 20$

62. $\dfrac{5x}{8} + \dfrac{y}{2} = 1$

$\dfrac{x}{2} + \dfrac{2y}{5} = 4$

SYSTEMS OF EQUATIONS AND INEQUALITIES

4.3   Solving Systems of Linear Equations by Elimination

Objective 1   Solve linear systems by elimination.

Solve each system by the elimination method. Check your answers.

1. $x + y = 5$
   $x - y = -3$

2. $3x - y = 5$
   $2x + y = 0$

3. $x - 4y = -4$
   $-x + y = -5$

4. $2x - y = 10$
   $3x + y = 10$

5. $4x + 3y = -4$
   $2x - 3y = 16$

6. $5x - 2y = 13$
   $2x + y = 7$

7. $8x + 2y = 14$
   $3x - 2y = -14$

8. $-2x + 5y = -8$
   $2x + y = -4$

9. $x - 3y = 5$
   $-x + 4y = -5$

10. $3x + 5y = 6$
    $x + 2y = 3$

11. $5x + 8y = 12$
    $3x - 8y = 20$

12. $3x + 2y = -19$
    $-3x + y = 22$

13. $5x - 6y = 9$
    $3x + 6y = 7$

14. $-x + 2y = -10$
    $x + 3y = -10$

15. $15x - 3y = 8$
    $21x + 3y = 10$

Objective 2   Multiply when using elimination method.

Solve each system by the elimination method. Check your answers.

16. $6x + 7y = 10$
    $2x - 3y = 14$

17. $8x + 6y = 10$
    $4x - y = 1$

18. $3x + 2y = 5$
    $2x - 3y = 12$

19. $3x + 5y = 8$
    $2x - y = -12$

20. $x - 4y = 10$
    $x + 6y = -10$

21. $6x + y = 1$
    $3x - 4y = 23$

22. $2x + y = 6$
    $-3x + y = -19$

23. $4x - 5y = -22$
    $3x + 2y = -5$

24. $4x - 9y = 7$
    $3x + 2y = 14$

25. $3x + y = 7$
    $2x + 3y = -7$

26. $3x - 7y = 12$
    $5x + 3y = -2$

27. $3x - 5y = -9$
    $2x + 3y = 13$

28. $4x - 3y = 7$
    $5x + 4y = 1$

29. $3x - 4y = 16$
    $4x + 5y = -20$

30. $3x + 2y = 16$
    $4x - 3y = -7$

Objective 3   **Use an alternative method to find the second value in a solution.**

Solve each system.

31. $5x - 3y = 23$
    $10 + 2y = 2x$

32. $4y = 2x - 2$
    $-9 + 3y = 5x$

33. $4x - 3y - 20 = 0$
    $6x + 5y + 8 = 0$

34. $5x + 5y + 15 = 0$
    $3x + 4y = -8$

35. $6x = 16 - 7y$
    $4x = 3y + 26$

36. $2x = 14 + 4y$
    $6y = -5x + 3$

37. $7 - y = 2x$
    $4x = 19 + 3y$

38. $5x + 3y + 4 = 0$
    $4x + 5y - 2 = 0$

39. $2x = 21 - 3y$
    $\frac{1}{3}x + \frac{2}{5}y = 3$

40. $\frac{1}{5}x + 3y = 46$
    $2x = 7 + \frac{1}{5}y$

41. $3x + 1 = 2y$
    $2y - 2 = 2x$

42. $2x + 5y = 11$
    $3x - 4y = -18$

Objective 4   **Use elimination to solve special systems.**

Solve each system by the elimination method.

43. $12x - 8y = 3$
    $6x - 4y = 6$

44. $2x + 4y = -6$
    $-x - 2y = 3$

45. $6x - 12y = 3$
    $2x - 4y = 1$

46. $6x - 2y = -4$
    $3x - y = -2$

47. $4x - 2y = -8$
    $2x - y = 4$

48. $36x + 20y = 8$
    $-27x - 15y = 2$

49. $15x + 6y = 9$
    $10x + 4y = 6$

50. $48x - 56y = 32$
    $21y - 18x = -12$

51. $12x - 18y = 10$
    $4x - 6y = 2$

52. $72x - 60y = -12$
    $25y - 30x = 5$

53. $x - 3y = 12$
    $3x - 9y = 4$

54. $15x - 10y = 6$
    $-12x + 8y = 2$

4.3 Mixed Exercises

Solve each system by the addition method. Check each solution.

55. $x + y = 6$
    $x - y = 2$

56. $2x - y = -9$
    $3x + y = 4$

57. $3x + 4y = 0$
    $x - 3y = -13$

58. $15x + 7y = -8$
    $-3x + 5y = 8$

59. $13x - 39y = 0$
    $x = 3y$

60. $9x - 15y = 21$
    $-12x + 20y = 28$

61. $4x - 5y = -1$
$2x + 3y = -17$

62. $4x = -2y - 10$
$7y - 1 + 5x = 0$

63. $6x - 5y = -12$
$9x + 2y = 1$

64. $3x = -9 - 3y$
$2x = 5 + y$

65. $6x = 2y - 8$
$5y = 20 + 15x$

66. $5x + 7y = -1$
$-5x + 13y = 16$

SYSTEMS OF EQUATIONS AND INEQUALITIES

4.4 Applications of Linear Systems

Objective 1 Solve problems about unknown numbers.

Use a system of equations to solve each problem.

1. The sum of two numbers is 64. Their difference is 18. Find the numbers.

2. Find two numbers whose sum is –66 and whose difference is –116.

3. The difference between two numbers is 27. If the larger is four more than twice the smaller, find the numbers.

4. The sum of two numbers is 20. Three times the smaller is equal to twice the larger. Find the numbers.

5. The difference between two numbers is 14. If two times the smaller is added to one-half the larger, the result is 52. Find the numbers.

6. Two towns have a combined population of 9045. There are 2249 more people living in one than in the other. Find the population in each town.

7. There are a total of 49 students in the two second grade classes at Jefferson School. If Carla has 7 more students in her class than Linda, find the number of students in each class.

8. A rope 82 centimeters long is cut into two pieces with one piece four more than twice as long as the other. Find the length of each piece.

9. The perimeter of a rectangular room is 50 feet. The length is three feet greater than the width. Find the dimensions of the rectangle.

10. The perimeter of a triangular pennant is 116 centimeters. If two sides are of equal length, and the third side is 20 centimeters longer than each of the equal sides, what are the lengths of the three sides.

Objective 2 Solve problems about quantities and their costs.

Use a system of equations to solve each problem.

11. Admission prices at a football game were $12 for adults and $9 for children. The total receipts for the game were $87,000. Tickets were sold to 8000 people. How many adults and how many children attended the game?

12. The receipts from a concert were $2100. The price for a regular ticket was $6 and the student tickets were half the regular price. If 400 tickets were sold, how many of each type were there?

13. Charlie has only $5 bills and $20 bills and has a total of $130. If there is a total of 11 bills, how many of each type are there?

14. A postal clerk has 1250 stamps in his drawer that are worth a total of $287.50. If there are only 25-cent and 15-cent stamps, how many of each kind are there?

15. There were 411 tickets sold for a soccer game, some for students and some for nonstudents. Student tickets cost $4.25 and nonstudent tickets cost $8.50 each. The total receipts were $3021.75. How many of each type were sold?

16. A cashier has some $5 bills and some $10 bills. The total value of the money is $750. If the number of tens is equal to twice the number of fives, how many of each type are there?

17. The total receipts for a basketball game were $4690.50. There were 723 tickets sold, some for children and some for adults. If the adult tickets cost $9.50 and the children's tickets cost $4, how many of each type were there?

18. Twice as many general admission tickets to a basketball game were sold as reserved seat tickets. General admission tickets cost $10 and reserved seat tickets cost $15. If the total value of both kinds of tickets was $26,250, how many tickets of each kind were sold?

19. Faustino has $10,000 to invest, part at 7% and part at 4%. If the total annual income from simple interest is to be $580, how much should he invest at each rate?

20. Carla has $12,000 to invest at 7% and 9%. She wants the income from simple interest on the two investments to total $1000 yearly. How much should she invest at each rate?

Objective 3   Solve problems about mixture.

Use a system of equations to solve each problem.

21. Steve wishes to mix coffee worth $6 a pound with coffee worth $9 a pound to get 45 pounds of a mixture worth $8 a pound. How many pounds of the $6 and the $9 coffee will be needed?

22. Ben wishes to blend candy selling for $1.60 a pound with candy selling for $2.50 a pound to get a mixture that will be sold for $1.90 a pound. How many pounds of the $1.60 and the $2.50 candy should be used to get 30 pounds of the mixture?

23. How many bags of coffee worth $90 a bag must be mixed with coffee worth $75 a bag to get 50 bags worth $87 a bag?

24. How many pounds of walnuts that sell for $8 a pound should be mixed with peanuts that sell for $6 a pound to produce a 10-pound mix that sells for $7 a pound?

25. Nina sells caramels that cost $3.65 per pound mixed with creams that cost $3.25 per pound. How much of each kind of candy is in a pound of the mixture if it costs $3.49?

26. How many liters of 75% solution should be mixed with a 55% solution to get 70 liters of 63% solution? How many liters of the 75% and 55% solutions should be used?

27. A 90% antifreeze solution is to be mixed with a 75% solution ot make 30 liters of an 80% solution. How many liters of the 90% and 75% solutions should be used?

28. Milton needs 45 liters of 20% solution. He has only 15% alcohol solution and 30% alcohol solution on hand to make the mixture. How many liters of each solution should he combine to make the mixture?

29. A 10% solution of antifreeze is mixed with a 50% solution of antifreeze to get 100 liters of a 22% solution. How many liters of each solution are used?

30. A 20% alcohol solution is to be mixed with a 5% solution to get 15 liters of a 10% solution. How many liters of each solution are used?

Objective 4   **Solve problems about distances, rate (or speed), and time.**

Use a system of equations to solve each problem.

31. Rick and Hilary drive from positions 378 miles apart and race toward each other. They meet after 3 hours. Find the average speed of each if Hilary travels 30 miles per hour faster than Rick.

32. Two trains start from positions 1242 miles apart and travel toward each other. They meet after $4\frac{1}{2}$ hours. Find the average speed of each train if one train travels 20 miles per hour faster than the other.

33. At the beginning of a fund-raising walk, Steve and Vic are 30 miles apart. If they leave at the same time and walk in the same direction, Steve would overtake Vic in 15 hours. If they walked toward each other, they would meet in 3 hours. What are their speeds?

34. Two bicyclists leave from Washington DC and ride in opposite directions. One travels $1\frac{1}{2}$ times as fast as the other. After 2 hours, they are 40 miles apart. Find the speed of each bicyclist.

35. Pablo left Somerset traveling to Akron 240 miles away at the same time as Faustino left Akron traveling to Somerset. They met after 2 hours. If Faustino was traveling twice as fast as Pablo, what were their speeds?

36. John left Louisville at noon on the same day that Mike left Louisville at 1 P.M. Both were traveling in the same direction. At 5 P.M., Mike was 62 miles behind John. If John was traveling 2 miles per hour faster than Mike, what were their speeds?

37. It takes a canoe 2 hours to go 16 miles downstream and 8 hours to return. Find the rate of the canoe in still water and the rate of the current.

38. A plane can travel 300 miles per hour with the wind and 230 miles per hour against the wind. Find the speed of the wind and the speed of the plane in still air.

39. It takes a kayak $1\frac{1}{2}$ hours to go 24 miles downstream and 4 hours to return. Find the speed of the current and the speed of the kayak in still water.

40. Two planes left Philadelphia traveling in opposite directions. Plane A left 15 minutes before plane B. After plane B had been flying for 1 hour, the planes were 860 miles apart. What were the speeds of the two planes if plane A was flying 40 miles per hour faster than plane B?

4.4 Mixed Exercises

Use a system of equations to solve each problem.

41. The perimeter of a rectangle is 144 centimeters. The length is 14 centimeters more than the width. Find the dimensions of the rectangle.

42. Luke plans to buy 10 ties with exactly $162. If some ties cost $14, and the others cost $25, how many ties of each price should he buy?

43. Stan has 14 bills in his wallet worth $95 altogether. If the wallet contains only $5 and $10 bills, how many bills of each denomination does he have?

44. Bill and Monica start in Washington and fly in opposite directions. At the end of 4 hours, they are 4896 kilometers apart. If Bill flies 60 kilometers per hour faster than Monica, what are their speeds?

45. A 30% acid solution is to be mixed with a 50% acid solution to get 100 milliliters of a 35% acid solution. How many milliliters of 30% solution are needed?

46. Sheila invested $50,000 in two accounts last year. One account earned 6% and the other earned 7% annual interest. If the total income last year from these accounts was $3350, how much was invested in each account?

47. It takes Carla's boat $\frac{1}{2}$ hour to go 8 miles downstream and 1 hour to make the return trip upstream. Find the speed of the current and the speed of Carla's boat in still water.

48. Hyman wishes to make 150 pounds of coffee blend that can be sold for $4 per pound. The blend will be a mixture of coffee worth $6 per pound and coffee worth $3 per pound. How many pounds of each kind of coffee should be used in the mixture?

49. Enid leaves Cherry Hill, driving by car toward New York, which is 186 miles away. At the same time, Hyman, riding his bicycle, leaves New York cycling toward Cherry Hill. Enid is traveling 28 miles per hour faster than Hyman. They pass each other $1\frac{1}{2}$ hours later. What are their speeds?

50. How many liters of water should be added to 25% antifreeze solution to get 30 liters of a 20% solution?

SYSTEMS OF EQUATIONS AND INEQUALITIES

4.5 Solving Systems of Linear Inequalities

Objective 1 Solve systems of linear inequalities by graphing.

Graph the solution of each system of linear inequalities.

1. $7x + 3y \geq 21$
   $x - y \leq 6$

2. $3x - y \leq 3$
   $x + y \leq 0$

3. $3x - y \leq 6$
   $3y - 6 \leq 2x$

4. $3x + 5y \geq 15$
   $y \geq x - 2$

5. $x + y \leq 3$
   $5x - y \geq 5$

6. $x + 2y \geq -4$
   $5x \leq 10 - 2y$

7. $3x - y > 3$
   $4x + 3y < 12$

8. $4x + 5y \leq 20$
   $y \leq x + 3$

9. $2x - y \geq 4$
   $5y + 15 \geq -3x$

10. $3x - 2y < 8$
    $x < 4$

11. $x - y \leq 2$
    $y \leq 2$

12. $x + y \geq 3$
    $x - 2y \leq 4$

13. $x < 2y + 3$
    $0 < x + y$

14. $6x - y > 6$
    $2x + 5y < 10$

15. $3x - 4y < 12$
    $y > -4$

16. $x - 2y \leq 4$
    $x + 2y \leq 4$

17. $y > 2$
    $4x - 3y < 9$

18. $4x - 3y > 12$
    $x < 4$

19. $4x - y \leq 4$
    $7y + 14 \geq -2x$

20. $5x - 2y \leq 10$
    $y \leq -2$

Chapter 5

EXPONENTS AND POLYNOMIALS

5.1 Adding and Subtracting Polynomials

Objective 1   Review combining like terms.

In each polynomial, add like terms whenever possible. Write the result in descending powers of the variable.

1. $-6s^3 + 12s^3$

2. $2t^6 + \left(-9t^6\right)$

3. $7z^3 - 4z^3 + 5z^3 - 11z^3$

4. $4 - 2x + 2x^2 - 7x^3 + 8x$

5. $.3m^5 - .7m^5$

6. $4.2r^2 - 5.7r - 2.5r - 4.9r^2$

7. $6c^3 - 9c^2 - 2c^2 + 14 + 3c^2 - 6c - 8 + 2c^3$

8. $4y^4 - 7y^2 + 4 - 7y^3 + 9y^4 - 2y^2 - 3y$

9. $4q^3 - q^4 + q^4 - 6q^2 - 8q^3$

10. $-12x^3 - 2x^2 + 4x - 7x^3$

11. $-\frac{1}{2}r^3 + \frac{1}{3}r + \frac{1}{4}r^3 - \frac{1}{3}r$

12. $\frac{4}{5}m^3 - \frac{2}{3}m^2 + \frac{1}{10}m^3 - \frac{1}{3}m^2$

Objective 2   Know the vocabulary for polynomials.

Choose one or more of the following descriptions for each expression: (a) polynomial,   (b) polynomial written in descending order, (c) not a polynomial.

13. $-2w^3 + 9w^2 + 4w - 10$

14. $x^4 + x^9$

15. $7y^4 - 3y^2 + \frac{2}{y}$

16. $3a^5 - 4a^3 - 3a^{-1} + 5$

17. $j^{-2} - 4j^{-1}$

18. $4k^6 - 7k^4 + 3k^3 - 2k + 5$

For each polynomial, first simplify, if possible, and write the resulting polynomial in descending powers of the variable. Then give the degree of this polynomial, and tell whether it is a monomial, a binomial, a trinomial, or none of these.

19. $-m^2 - 4m + 6$

20. $7m^2 + 3m + m^4$

21. $3z^4 + 5z^3 - 2z^3$

22. $2p^3 - 4p^5 - 2p^5$

23. $3n^8 - n^2 - 2n^8$

24. $4r^4 + 3r^2 - 7r^4$

25. $\frac{7}{8}x^2 - \frac{3}{4}x - \frac{3}{8}x^2 + \frac{1}{4}x$

26. $\frac{5}{6}y^2 + \frac{1}{3}y + \frac{1}{6} - \frac{1}{3}y^2 + \frac{2}{3} - \frac{5}{6}y$

Objective 3   Evaluate polynomials

Objective 4   Add polynomials.

Add each pair of polynomials.

27. $5m^4 + 2m^3 - 4$
$\underline{-3m^4 + 5m^3 - 3}$

28. $7y^3 - 3y^2 + 3$
$\underline{-2y^3 - 3y^2 + 10}$

29. $9m^3 + 4m^2 - 2m + 3$
$\underline{-4m^3 - 6m^2 - 2m + 1}$

30. $5w^4 + 2m^2 - 6m + 6$
$\underline{2w^4 - 2m^2 + 7m + 8}$

31. $5x^5 - 2x^2 - 5x - 1$
$\underline{-6x^5 \qquad + 7x + 3}$

32. $9y^3 \qquad - 4y + 7$
$\underline{\qquad -9y^2 + 4y + 4}$

33. $7p^4 \qquad + 5p^2 - 2p + 8$
$\underline{2p^4 - 3p^3 - 8p^2 - 3p + 14}$

34. $8z^5 - 7z^4 + 9z^3 - 5z^2 + z + 4$
$\underline{-8z^5 - 2z^4 - 9z^3 + 2z^2 + z \qquad}$

35. $(3x^2 + 2x^4 - 3) + (8x^3 - 5x^4 - 6x^2)$

36. $(9x^3 - 5x^2 - 2x + 4) + (7x^3 + 3x^2 - 3)$

37. $(7y^4 + 2y^3 + 2y - 4) + (2y^3 - 3y + 1)$

38. $(-4a^5 - 6a^3 + 5a + 2) + (4a^3 - 7a - 6)$

39. $(x^5 + 2x^3 - x^2) + (3x^5 - 7x^4 - 2x^3 + 4x^2 - x)$

40. $(m^4 - 3m^3 + 6m^2 + m + 3) + (m^3 + 3m^2 - m - 7)$

Objective 5   Subtract polynomials.

Subtract.

41. $9x^2 + 2x$
$\underline{3x^2 + 4x}$

42. $6m^3 - 5m$
$\underline{7m^3 - 3m}$

43. $2m^2 - 5m + 1$
$\underline{-2m^2 - 5m + 3}$

44. $10k^2 - 2k - 8$
$\underline{-6k^2 - 4k + 14}$

45. $6n^3 + 3n + 1$
$\underline{7n^3 - 3n - 2}$

46. $2x^5 - 4x^3 + 8x$
$\underline{3x^5 + 2x^3 - 3x}$

47.
$$8y^4 - 3y^3 + 2y^2 - 2y - 1$$
$$\underline{-11y^4 + 5y^3 - 2y^2 \qquad -1}$$

48.
$$2z^5 + z^4 - 2z^3 \qquad\quad + 5$$
$$\underline{\quad 2z^4 \qquad -7z^2 + 8z - 2}$$

49. $\left(9x^3 + 7x^2 - 6x + 3\right) - \left(6x^3 - 6x + 1\right)$

50. $\left(2x^3 - 4x^2 + 3x + 10\right) - \left(6x^3 - 4x + 2\right)$

51. $\left(8p^2 + 7p - 2\right) - \left(3p^2 - 3p + 7\right) - \left(2p^2 - 3\right)$

52. $\left(4z^4 - 2z^2 + 4z\right) - \left(3z^3 - 2z^2 - 4z - 7\right)$

53. $\left(2a^5 - 6a^2 - 2a + 9\right) - \left(-3a^5 + 2a^3 - 9a^2 - 2a - 4\right)$

54. $\left(p^4 - 9p^3 - 2p^2 + 7p - 4\right) - \left(-2p^4 - 4p^2 + 3\right)$

Objective 6 **Apply the rules and definitions for polynomials to multivariable polynomials.**

Add or subtract as indicated. Give the degree of the answer.

55. $\left(4x^2y + 5xy + 7y^2\right) + \left(-2x^2y - 5xy + 3y^2\right)$

56. $\left(-2a^6 + 8a^4b - b^2\right) - \left(a^6 + 7a^4b + 2b^2\right)$

57. $\left(-4m^2n + 3n - 6m\right) - \left(2m + 7n + 4nm^2\right)$

58. $\left(4ab + 2bc - 9ac\right) + \left(3ca - 2cb - 9ba\right)$

59. $\left(2x^2y + 2xy - 4xy^2\right) + \left(6xy + 9xy^2\right) - \left(9x^2y + 5xy\right)$

60. $\left(.01ab + .03a^2 - .05b^2\right) - \left(-.08a^2 + .02b^2 + .01ab\right)$

Add.

61.
$$5x^3y - 3x^2y^2 + 5xy^3$$
$$\underline{-3x^3y + 3x^2y^2 - 7xy^3}$$

62.
$$6c^3d + 4c^2d - 7d^2$$
$$\underline{-8c^3d + 3c^2d - 2d^2}$$

Subtract.

63.
$$13a^3b \qquad\quad + 2ab^3 + 7$$
$$\underline{16a^3b - 6a^2b^2 - 7ab^3 - 4}$$

64.
$$-6rs + 2rt - 5st$$
$$\underline{\quad 5rs - 7rt - \ st}$$

5.1 Mixed Exercises

Perform the indicated operations. Write each resulting polynomial in descending powers of the variable. Then give the degree of this polynomial, and tell whether it is a **monomial,** *a* **binomial,** *a* **trinomial,** *or* **none of these.**

65. $\left(b^3 + 4b^2 + 7b + 2\right) + \left(8b^3 + 6b + 3\right)$ 66. $\left(2y^2 + 2y - 5\right) - \left(4y^2 - 2y + 7\right)$

67. $\left(2y^4 + 3y^3 + 7y - 3\right) + \left(2y^3 + 4y + 5\right)$ 68. $\left(6m^3 - 3m + 2\right) - \left(-5m^2 - 8\right)$

69. $\left(x^5 + 2x^3 - x^2\right) + \left(4x^5 - 3x^4 - 2x^3 + x^2 - 2x\right)$

70. $\left(2p^4 - 5p^3 - 3p^2 + 8p - 2\right) - \left(-2p^4 - 3p^2 + 2\right)$

71. $10 - \left(4 - 2y - 11y^2\right)$

72. $\left(k^2 - 2k + 8\right) + \left(-3k^2 - 4k - 1\right) + \left(2k^2 + 6k - 6\right)$

73. $\left(5x^3 + x - 3\right) - \left(2x^3 - x - 5\right) + \left(x^3 - 2x - 7\right)$

74. $\left(4m^2 - 3\right) + \left(2m^2 + m\right) - \left(2m^2 + 9\right)$

EXPONENTS AND POLYNOMIALS

5.2 The Product Rule and Power Rules for Exponents

Objective 1 Use Exponents.

Write each expression in exponential form and evaluate.

1. $2 \cdot 2 \cdot 2 \cdot 2 \cdot 2 \cdot 2$

2. $10 \cdot 10 \cdot 10$

3. $(-1)(-1)(-1)(-1)(-1)(-1)(-1)(-1)$

4. $(.4)(.4)(.4)$

5. $\left(\frac{1}{3}\right)\left(\frac{1}{3}\right)\left(\frac{1}{3}\right)\left(\frac{1}{3}\right)\left(\frac{1}{3}\right)$

6. $\left(-\frac{2}{5}\right)\left(-\frac{2}{5}\right)\left(-\frac{2}{5}\right)\left(-\frac{2}{5}\right)$

Write each expression in exponential form.

7. $r \cdot r \cdot r \cdot r \cdot r$

8. $(-2y)(-2y)(-2y)(-2y)(-2y)$

9. $\left(\frac{1}{4}mn\right)\left(\frac{1}{4}mn\right)\left(\frac{1}{4}mn\right)\left(\frac{1}{4}mn\right)$

10. $(.5st)(.5st)(.5st)(.5st)$

Evaluate each exponential expression. Name the base and the exponent.

11. $(-4)^4$

12. -3^4

13. 2^6

14. $(-5)^4$

15. -6^2

16. $\left(-\frac{1}{3}\right)^4$

Objective 2 Use the product rule for exponents.

Use the product rule to simplify each expression, if possible. Write each answer in exponential form.

17. $4^7 \cdot 4^4$

18. $(-3)^4 \cdot (-3)^9$

19. $(.1)^7 (.1)^5 (.1)^2$

20. $3^2 \cdot 5^2$

21. $\left(\frac{1}{2}\right)^9 \cdot \left(\frac{1}{2}\right)$

22. $\left(\frac{2}{5}\right)^6 \cdot \left(\frac{2}{5}\right)^4$

23. $(-5)^{10} \cdot (-5)^4$

24. $10^2 + 10^3$

Multiply.

25. $9p^4 \cdot 6p^5$

26. $10n^4 \cdot 10n \cdot n^7$

27. $7a \cdot 12a^7$

28. $12b \cdot \left(-12b^{11}\right)$ **29.** $\left(-4r^2\right)\left(-2r^4\right)$ **30.** $\left(-4x^3\right)\left(8x^{12}\right)$

In each of the following exercises, first add the given terms. Then start over and multiply them.

31. $7y^2, 9y^2$ **32.** $-3m^6, \ -4m^6$

33. $-5a^3, \ 4a^3, \ -a^3$ **34.** $4t^2, \ 6t^2, \ -2t^2$

Objective 3   **Use the rule** $\left(a^m\right)^n = a^{mn}$.

Simplify each expression. Write all answers in exponential form.

35. $\left(3^4\right)^3$ **36.** $\left(7^3\right)^9$ **37.** $\left(6^4\right)^6$ **38.** $\left[\left(\frac{1}{3}\right)^3\right]^5$

39. $\left(8^5\right)^7$ **40.** $\left(9^5\right)^9$ **41.** $\left[\left(-3\right)^3\right]^7$ **42.** $\left[\left(-2\right)^7\right]^5$

43. $-\left(13^3\right)^8$ **44.** $\left(14^5\right)^5$ **45.** $\left(6^{11}\right)^{10}$ **46.** $-\left(21^5\right)^4$

Objective 4   **Use the rule** $\left(ab\right)^m = a^m b^m$.

Simplify each expression.

47. $\left(\frac{1}{3}x^4\right)^2$ **48.** $\left(yz^4\right)^3$ **49.** $\left(p^2q^3\right)^4$ **50.** $\left(r^7s^2\right)^2$

51. $5\left(ab^3\right)^3$ **52.** $\left(2w^3z^7\right)^4$ **53.** $\left(5r^3t^2\right)^4$ **54.** $\left(-.2a^4b\right)^3$

55. $\left(-2r^3s\right)^6$ **56.** $2\left(3c^4d^7\right)^4$ **57.** $-2\left(3x\right)^4$ **58.** $\left(4c^3d^4\right)^3$

Objective 5   **Use the rule** $\left(\dfrac{a}{b}\right)^m = \dfrac{a^m}{b^m}$.

Simplify each expression. Assume all variables represent nonzero real numbers.

59. $\left(\dfrac{3}{5}\right)^3$ **60.** $\left(\dfrac{5}{8}\right)^2$ **61.** $\left(\dfrac{-2a}{b^2}\right)^7$ **62.** $\left(\dfrac{w}{2}\right)^4$

63. $\left(\dfrac{z^2}{3}\right)^3$ **64.** $\left(\dfrac{-2x}{5}\right)^2$ **65.** $\left(\dfrac{xy}{z^2}\right)^4$ **66.** $\left(\dfrac{x^3y}{z^4}\right)^9$

67. $\left(-\dfrac{2x}{5}\right)^3$ **68.** $\left(\dfrac{2a^2}{b^3}\right)^5$ **69.** $\left(\dfrac{x}{2y}\right)^4$ **70.** $\left(\dfrac{4}{g}\right)^3$

Objective 6   Use combinations of the rules for exponents.

Simplify. Write all answers in exponential form.

71. $(7a)^4 (7a)^5$ **72.** $(-4q)^3 (-4q)^6$ **73.** $\left(\dfrac{4^2}{7^3}\right)^3 \cdot 4^7$

74. $\left(\dfrac{3b^2}{11}\right)^4$ **75.** $\left(\dfrac{3}{4}\right)^3 (5x)^4$ **76.** $\left(\dfrac{1}{3}\right)^3 (4ab^2)^3$

77. $(-x^3)^2 (-x^5)^4$ **78.** $(5x^2 y^3)^7 (5xy^4)^4$ **79.** $\left(\dfrac{7a^2 b^3}{2}\right)^7$

80. $\left(\dfrac{km^4 p^2}{3n^4}\right)^7 \;(n \neq 0)$ **81.** $(-2pq)^3 (-2pq)^6$ **82.** $(2ab^2 c)^5 (ab)^4$

Objective 7   Use the rules for exponents in an application for geometry.

5.2 Mixed Exercises

Evaluate each exponential expression. Name the base and the exponent.

83. $\left(-\tfrac{1}{3}\right)^4$ **84.** -2^6 **85.** $(-1)^5$

86. -1^7 **87.** -5^3 **88.** $(-2)^4$

Simplify each expression. Write all answers in exponential form.

89. $\left(6^4\right)^3$ **90.** $\left(x^4 y\right)^7$ **91.** $\left(3a^7 b^3\right)^4$

92. $\left(\dfrac{3x}{5y}\right)^8 \;(y \neq 0)$ **93.** $\left(3m^3 n^4\right)^4$ **94.** $(4wt)^3 (4wt)^9$

95. $(-4a^4)(-4a^3)^4$ **96.** $\left(7a^2 b^2 c\right)^3 \left(ab^3 c^4\right)^4$ **97.** $\left(z^3\right)^7 \left(z^2\right)^3$

98. $\left(-\dfrac{3}{4}\right)^2 \left(3a^2\right)^4$

99. $\left(\dfrac{7wx^5}{y^2}\right)^4 \quad (y \neq 0)$

100. $\left(\dfrac{r^3 st}{2n^2}\right)^4 \quad (n \neq 0)$

EXPONENTS AND POLYNOMIALS

5.3 Multiplying Polynomials

Objective 1 Multiply a monomial and a polynomial.

Find each product.

1. $(3y^3)(4y^2)$

2. $(-2y^3)(-8y^4)$

3. $7z(5z^3+2)$

4. $-2x^4(3+6x+2x^2)$

5. $-6z(z^5+3z^3+4z+2)$

6. $-3y^2(2y^3+3y^2-4y+11)$

7. $4k^2(3+2k^3+6k^4)$

8. $2m(3+7m^2+3m^3)$

9. $7b^2(-5b^2+1-4b)$

10. $-4r^4(2r^2-3r+2)$

11. $8mn(4m^2+2mn+7n^2)$

12. $-3r^2s^3(8r^2s^2-4rs+2rs^2)$

Objective 2 Multiply two polynomials.

Find the product.

13. $(x+3)(x+9)$

14. $(y-2)(y-4)$

15. $(3p+4)(p+2)$

16. $(7+2a)(4+3a)$

17. $(5n+1)(2n+5)$

18. $(2x+5)(3x+4)$

19. $(3m-5)(2m+4)$

20. $(x+3)(x^2-3x+9)$

21. $(y+4)(y^2-4y+16)$

22. $(r+3)(2r^2-3r+5)$

23. $(3y-4)(3y^3-2y^2-y+4)$

24. $(2m^2+1)(3m^3+2m^2-4m)$

25. $(2z^2-3)(z^4+2z^3+z^2+3z+2)$

26. $(2x^2+3x+3)(2x^2-3x+4)$

Find each product, using the vertical method of multiplication.

27. $(2x+3)(2x^2-3x+2)$

28. $(2y-3)(3y^3+2y^2-y+2)$

29. $(2a^2+1)(3a^3-2a^2+a)$

30. $(4b^2-3)(b^4-2b^3+b^2-b-2)$

31. $(2x^2+3x+2)(4x^3+2x+3)$

32. $(3y^3-y^2+2)(-3y^3-3y+2)$

33. $(a+5)(a+5)$

34. $(2r+s)(2r+s)$

35. $(2m^3+3m-3)(-2m^3-4m+1)$

36. $(3x^2+x)(2x^2+3x-4)$

37. $(2x^2-x+2)(2x^2-x+2)$

38. $(y^2-2y-3)(y^2-2y-3)$

Objective 3   Multiply binomials by the FOIL method.

Use the FOIL method to find each product.

39. $(4m+3)(m-7)$

40. $(3x+2y)(2x-3y)$

41. $(5a-b)(4a+3b)$

42. $(11k-4)(11k+4)$

43. $(3+4a)(1+2a)$

44. $(4x-3y)(x+2y)$

45. $(2m+3n)(-3m+4n)$

46. $(2v^2+w^2)(v^2-3w^2)$

47. $(x-.5)(x+.3)$

48. $(2y+.1)(2y-.5)$

49. $(x-\frac{1}{3})(x-\frac{4}{3})$

50. $(z+\frac{4}{5})(z-\frac{2}{5})$

5.3   Mixed Exercises

Find each product.

51. $(7x^3)(4x^4)$

52. $-5r^4(2r^3-4r^2+3)$

53. $(a+4)(a-2)$

54. $(9p+5r)(3p-2r)$

55. $(3t+2u)(5t-2u)$

56. $3y^2(6y^3-2y^2-3y+4)$

57. $(x-4y)(x+4y)$

58. $-6p^4(-7+4p-2p^3+4p^5)$

59. $(x+2)(3x^3-2x^2+4)$

60. $(3x^2-2x-3)(x-4)$

61. $\left(x^2-4x+3\right)\left(x^2-4x+3\right)$

62. $\left(2x^3-5x^2+4x-3\right)\left(x^2+2x-1\right)$

63. $\left(3x^3+3x^2-2x+1\right)\left(2x^2-x+2\right)$

EXPONENTS AND POLYNOMIALS

5.4 Special Products

Objective 1 Square binomials.

Find each square by using the pattern for the square of a binomial.

1. $(z+3)^2$

2. $(t+4)^2$

3. $(a+2b)^2$

4. $(5y-3)^2$

5. $(2m+5)^2$

6. $(5m+3n)^2$

7. $(7+x)^2$

8. $(5+2y)^2$

9. $(2p+3q)^2$

10. $(2m-3p)^2$

11. $(4y-.7)^2$

12. $(4x-\frac{1}{4}y)^2$

13. $(3x-\frac{1}{3}y)^2$

14. $(3a+\frac{1}{2}b)^2$

Objective 2 Find the product of the sum and difference of two terms.

Find each product by using the pattern for the sum and difference of two terms.

15. $(z-6)(z+6)$

16. $(12+x)(12-x)$

17. $(7x-3y)(7x+3y)$

18. $(8k+5p)(8k-5p)$

19. $(4p+7q)(4p-7q)$

20. $(2+3x)(2-3x)$

21. $(9-4y)(9+4y)$

22. $(x+.2)(x-.2)$

23. $(y+\frac{4}{3})(y-\frac{4}{3})$

24. $(7m-\frac{3}{4})(7m+\frac{3}{4})$

25. $(2a+\frac{4}{3}b)(2a-\frac{4}{3}b)$

26. $(\frac{3}{4}s+\frac{7}{5}t)(\frac{3}{4}s-\frac{7}{5}t)$

27. $(y^2+2)(y^2-2)$

28. $(5m^4-7n^3)(5m^4+7n^3)$

Objective 3 **Find higher powers of binomials.**

Find each product.

29. $(x+2)^3$

30. $(a-3)^3$

31. $(y+4)^3$

32. $(2x-3)^3$

33. $(2x+1)^3$

34. $(k+2)^4$

35. $(t-3)^4$

36. $(x+2y)^4$

37. $(3b-2)^4$

38. $(4s+3t)^4$

39. $(2x-1)^3$

40. $(j+3)^4$

5.4 Mixed Exercises

Find each product.

41. $(4y+5)^2$

42. $(n+3)(n-5)$

43. $(4b-9)(4b+9)$

44. $(3b-5)^2$

45. $(7t+6u)(7t-6u)$

46. $(3p-2)(p+4)$

47. $(7h-3k)(3h+2k)$

48. $(5k+3m)^2$

49. $(5-6w)(5+6w)$

50. $(8a-3)(4a+5)$

51. $\left(4x+\frac{7}{4}\right)\left(4x-\frac{7}{4}\right)$

52. $(2y-.3)^2$

53. $\left(4j+\frac{1}{2}k\right)^2$

54. $\left(\frac{4}{7}t+2u\right)\left(\frac{4}{7}t-2u\right)$

EXPONENTS AND POLYNOMIALS

5.5   Integer Exponents and the Quotient Rule

Objective 1   Use zero as an exponent.

Evaluate each expression.

1. 7^0

2. $-(-8)^0$

3. -12^0

4. $\left(\frac{7}{9}\right)^0$

5. $2^0 + 6^0$

6. $.2^0 + (-.2)^0$

7. $\left(\frac{2}{3}\right)^0 + \left(\frac{1}{3}\right)^0 - 2^0$

8. $(-5)^0 - (-5)^0$

9. $-12^0 + (-12)^0$

10. $-15^0 - (-15)^0$

11. $-r^0 \;\; (r \neq 0)$

12. $\dfrac{0^8}{8^0}$

Objective 2   Use negative numbers as exponents.

Evaluate each expression.

13. 4^{-2}

14. $\left(\frac{2}{7}\right)^{-1}$

15. $(-3)^{-3}$

16. $(-2)^{-4}$

17. $\left(\frac{3}{5}\right)^{-2}$

18. $\left(\frac{2}{9}\right)^{-1}$

19. $8^{-1} + 4^{-1}$

20. $10^{-2} + 5^{-2}$

Simplify by using the definition of negative exponents. Write each expression with only positive exponents. Assume all variables represent nonzero real numbers.

21. r^{-7}

22. a^{-4}

23. $\dfrac{2}{r^{-7}}$

24. $\dfrac{r^{-2}}{6}$

25. $\dfrac{2x^{-4}}{3y^{-7}}$

26. $\dfrac{2^{-4}}{8^{-2}}$

Objective 3   Use quotient rule for exponents.

Use the quotient rule to simplify each expression. Write answers with only positive exponents. Assume that all variables represent nonzero real numbers.

27. $\dfrac{2^9}{2^5}$

28. $\dfrac{9^{12}}{9^7}$

29. $\dfrac{(-2)^8}{(-2)^3}$

30. $\dfrac{(-4)^{-4}}{(-4)^{2}}$

31. $\dfrac{2^{4} \cdot x^{2}}{2^{5} \cdot x^{8}}$

32. $\dfrac{4k^{7}m^{10}}{8k^{3}m^{5}}$

33. $\dfrac{12x^{9}y^{5}}{12^{4}x^{3}y^{7}}$

34. $\dfrac{3^{-5}}{3^{-8}}$

35. $\dfrac{12^{-7}}{12^{-6}}$

36. $\dfrac{a^{4}b^{3}}{a^{-2}b^{-3}}$

37. $\dfrac{3^{-1}m^{-4}p^{6}}{3^{4}m^{-1}p^{-2}}$

38. $\dfrac{8b^{-3}c^{4}}{8^{-4}b^{-7}c^{-3}}$

Objective 4   Use combination of rules.

Simplify each expression. Write answers with only positive exponents. Assume that all variables represent nonzero real numbers.

39. $\dfrac{\left(7^{2}\right)^{6}}{7^{6}}$

40. $\dfrac{\left(9^{3}\right)^{2}}{9^{5}}$

41. $\dfrac{6^{-2} \cdot 6^{3}}{6^{-3}}$

42. $x^{-3} \cdot x^{3} \cdot x^{9}$

43. $\left(2^{-4}\right)^{4}$

44. $\left(3^{4}x^{-2}y\right)^{4}$

45. $\left(5w^{2}y^{2}\right)^{-2}\left(4wy^{-3}\right)^{2}$

46. $\left(2p^{-3}q^{2}\right)^{2}\left(4p^{4}q^{-1}\right)^{-1}$

47. $\dfrac{(2y)^{-4}}{(3y)^{-2}}$

48. $(9xy)^{7}(9xy)^{-8}$

49. $\dfrac{c^{10}\left(c^{2}\right)^{3}}{\left(c^{3}\right)^{3}\left(c^{2}\right)^{-9}}$

50. $\dfrac{\left(a^{-1}b^{-2}\right)^{-4}\left(ab^{2}\right)^{6}}{\left(a^{3}b\right)^{-2}}$

51. $\left(\dfrac{x^{8}y^{2}}{x^{6}y^{-4}}\right)^{-2}$

52. $\left(\dfrac{k^{3}t^{4}}{k^{2}t^{-1}}\right)^{-4}$

53. $\dfrac{\left(3^{-2}x^{-5}y\right)^{-4}\left(2x^{2}y^{-4}\right)^{2}}{\left(2x^{-2}y^{2}\right)^{-2}}$

54. $\dfrac{\left(4a^{-1}b^{4}\right)^{-3}\left(4a^{2}b\right)^{-1}}{\left(2a^{-3}b^{2}\right)^{-2}}$

5.5 Mixed Exercises

Evaluate each expression.

55. $(-4)^{0}+(8)^{0}$

56. $2^{2}+2^{-2}$

57. $-7^{0}+(-7)^{0}$

58. $\left(2^{-4}\right)^{0}$ **59.** $\left(3^{0}\right)^{-3}$ **60.** $\left(\frac{2}{3}\right)^{-3}$

Simplify each expression. Write answers with only positive exponents. Assume that all variables represent nonzero real numbers.

61. $\dfrac{7^{-7}}{7^{-9}}$

62. $\dfrac{a^{-8}}{a^{4}}$

63. $\dfrac{\left(b^{2}c^{3}\right)^{3}}{c^{4}}$

64. $\dfrac{8^{5}\cdot 8^{-4}}{8^{-2}}$

65. $\left(\dfrac{3}{4}\right)^{-2}\cdot\left(\dfrac{4}{3}\right)$

66. $\dfrac{\left(4k^{-1}\right)^{3}}{4k^{3}}$

67. $\dfrac{\left(2xy^{-1}\right)^{3}}{2^{3}x^{-3}y^{2}}$

68. $\dfrac{x^{3}\cdot x^{-3}}{x^{5}\cdot x^{-2}}$

69. $\left(\dfrac{3x^{2}}{2x^{-4}}\right)^{-1}$

70. $\dfrac{\left(r^{3}\right)^{-2}\left(s^{2}\right)^{4}}{r^{0}\left(s^{-3}\right)^{-2}}$

71. $\dfrac{\left(3y^{-1}z^{4}\right)^{-1}\left(3y^{-2}\right)}{\left(y^{3}z^{2}\right)^{-2}}$

72. $\dfrac{\left(2^{-1}m^{-1}n^{-1}p\right)^{-2}}{\left(p^{2}m^{-1}\right)^{-3}}$

EXPONENTS AND POLYNOMIALS

5.6   Dividing a Polynomial by a Monomial

Objective 1   Divide a polynomial by a Monomial.

Divide each polynomial by $4m^2$.

1. $16m^3 + 8m^2$

2. $32m^4 - 8m^3 + 12m^2$

3. $28m^5 - 4m^2$

4. $36m^6 + 24m^4 - 44m^2 - 8$

5. $28m^3 + 48m^2 - 16m$

6. $32m^3 - 2m^2$

7. $-84m^3 + 36m^2 + 8m$

8. $4m^2 - 4$

9. $32m^3 - 48m - 12$

10. $16m^5 - 20m^4 + 4m^2 - 2m + 3$

Perform each division.

11. $\dfrac{6p^4 + 18p^7}{6p^4}$

12. $\dfrac{12x^6 + 28x^5 + 20x^3}{4x^2}$

13. $\left(20a^3 - 9a\right) \div \left(4a\right)$

14. $\left(6z^5 + 27z^3 - 12z + 10\right) \div \left(3z\right)$

15. $\left(20x^4 - 10x^2\right) \div \left(2x\right)$

16. $\left(9x^4 + 24x^3 - 48x + 12\right) \div \left(3x\right)$

17. $\left(m^2 + 7m - 42\right) \div \left(2m\right)$

18. $\dfrac{40p^4 - 35p^3 - 15p}{5p^2}$

19. $\dfrac{70q^4 - 40q^2 + 10q}{10q^2}$

20. $\dfrac{2y^9 + 8y^6 - 41y^3 - 12}{y^3}$

21. $\dfrac{12z^5 + 28z^4 - 8z^3 + 3z}{4z^3}$

22. $\dfrac{48x + 64x^4 + 2x^8}{4x}$

23. $\dfrac{-25u^3v + 20u^2v^2 - 45uv^3}{5uv}$

24. $\dfrac{21y^2 - 14y + 42}{-7y^2}$

EXPONENTS AND POLYNOMIALS

5.7   Dividing a Polynomial by a Polynomial

Objective 1   Divide a polynomial by a polynomial.

Perform each division.

1. $\dfrac{x^2 - x - 6}{x - 3}$

2. $\dfrac{y^2 - 2y - 24}{y + 4}$

3. $\dfrac{18a^2 - 9a - 5}{3a + 1}$

4. $\dfrac{p^2 + 5p - 24}{p - 3}$

5. $\left(x^2 + 16x + 64\right) \div \left(x + 8\right)$

6. $\left(r^2 - 2r - 20\right) \div \left(r - 5\right)$

7. $\left(2a^2 - 11a + 16\right) \div \left(2a + 3\right)$

8. $\left(9w^2 + 12w + 4\right) \div \left(3w + 2\right)$

9. $\dfrac{5w^2 - 22w + 4}{w - 4}$

10. $\dfrac{5b^2 + 32b + 3}{b + 7}$

11. $\dfrac{9m^2 - 18m + 16}{3m - 4}$

12. $\dfrac{81a^2 - 1}{9a + 1}$

13. $\dfrac{4x^2 - 25}{2x - 5}$

14. $\dfrac{12y^3 - 11y^2 + 9y + 18}{4y + 3}$

15. $\dfrac{2z^3 - 7z^2 + 3z + 2}{2z + 3}$

16. $\dfrac{6m^3 + 7m^2 - 13m + 16}{3m + 2}$

17. $\left(27p^4 - 36p^3 - 6p^2 + 26p - 24\right) \div \left(3p - 4\right)$

18. $\left(3x^3 - 11x^2 + 25x - 25\right) \div \left(x^2 - 3x - 5\right)$

19. $\dfrac{6x^4 - 12x^3 + 13x^2 - 5x - 1}{2x^2 + 3}$

20. $\dfrac{12y^5 - 8y^4 - y^3 + 2y^2 - 5}{4y^2 - 3}$

21. $\dfrac{2a^4 + 5a^2 + 3}{2a^2 + 3}$

22. $\dfrac{3x^4 + 2x^3 - 2x^2 - 2x - 2}{x^2 - 1}$

23. $\dfrac{y^3 + 1}{y + 1}$

24. $\dfrac{b^4 - 1}{b^2 - 1}$

25. $\dfrac{6x^5 + 7x^4 - 7x^3 + 7x + 4}{3x + 2}$

26. $\dfrac{32x^5 - 243}{2x - 3}$

Objective 2   Apply division to a geometric problem.

EXPONENTS AND POLYNOMIALS

5.8 An Application of Exponents: Scientific Notation

Objective 1 Express numbers in scientific notation.

Write each number in scientific notation.

1. 325

2. 4579

3. 23,651

4. 209,907

5. 7.42

6. 429,600,000,000

7. .0257

8. .246

9. .00000413

10. .00426

11. −43,276

12. −.00047

Objective 2 Convert numbers in scientific notation to numbers without exponents.

Write each number without exponents.

13. 2.5×10^4

14. 7.2×10^7

15. -2.45×10^6

16, 4.045×10^0

17. 2.3×10^4

18. 4.5×10^7

19. 6.4×10^{-3}

20. 7.24×10^{-4}

21. 4.007×10^{-2}

22. 4.752×10^{-1}

23. -4.02×10^0

24. -9.11×10^{-4}

Objective 3 Use scientific notation in calculations.

Perform the indicated operations with the numbers in scientific notation, and then write your answers without exponents.

25. $\left(7 \times 10^7\right) \times \left(3 \times 10^0\right)$

26. $\left(3 \times 10^6\right) \times \left(4 \times 10^{-2}\right) \times \left(2 \times 10^{-1}\right)$

27. $\left(2.3 \times 10^4\right) \times \left(1.1 \times 10^{-2}\right)$

28. $\left(2.3 \times 10^{-4}\right) \times \left(3.1 \times 10^{-2}\right)$

29. $\dfrac{4.6 \times 10^{-3}}{2.3 \times 10^{-1}}$

30. $\dfrac{8.5 \times 10^{-3}}{1.7 \times 10^{-7}}$

31. $\dfrac{5.2 \times 10^4}{1.3 \times 10^{-2}}$

32. $\dfrac{9.39 \times 10^1}{3 \times 10^3}$

33. $\left(6 \times 10^4\right) \times \left(3 \times 10^5\right) \div \left(9 \times 10^7\right)$

34. $\left(3 \times 10^4\right) \times \left(4 \times 10^2\right) \div \left(2 \times 10^3\right)$

35. $\dfrac{\left(7.5\times10^6\right)\times\left(4.2\times10^{-5}\right)}{\left(6\times10^4\right)\times\left(2.5\times10^{-3}\right)}$

36. $\dfrac{\left(2.1\times10^{-3}\right)\times\left(4.8\times10^4\right)}{\left(1.6\times10^6\right)\times\left(7\times10^{-6}\right)}$

5.8 Mixed Exercises

Write each number in scientific notation.

37. 2,705,000

38. 46

39. .0253

40. .75

41. −4.327

42. −.0000403

Perform the indicated operations with the numbers in scientific notation, and then write your answers without exponents.

43. $\left(1.6\times10^5\right)\times\left(3.4\times10^{-3}\right)$

44. $\left(2.5\times10^{-7}\right)\times\left(5.6\times10^4\right)$

45. $\dfrac{6\times10^{-3}}{4\times10^{-7}}$

46. $\dfrac{5\times10^{-4}}{5\times10^{-5}}$

47. $\dfrac{1.26\times10^2}{6\times10^{-4}}$

48. $\dfrac{6\times10^{-4}}{2\times10^{-7}}$

49. $\dfrac{\left(6\times10^4\right)\times\left(3\times10^5\right)}{9\times10^7}$

50. $\dfrac{\left(8\times10^{-2}\right)\times\left(6\times10^3\right)}{1.2\times10^5}$

51. $\dfrac{\left(12.5\times10^3\right)\times\left(4.9\times10^{-4}\right)}{\left(7\times10^{-8}\right)\times\left(2.5\times10^5\right)}$

52. $\dfrac{\left(3.8\times10^6\right)\times\left(7.2\times10^{-5}\right)}{\left(9\times10^2\right)\times\left(1.9\times10^{-4}\right)}$

53. $\dfrac{\left(4\times10^{-5}\right)\times\left(3\times10^4\right)}{\left(6\times10^{-2}\right)\times\left(2\times10^{-1}\right)}$

54. $\dfrac{\left(8\times10^2\right)\times\left(3\times10^{-5}\right)}{\left(6\times10^3\right)\times\left(4\times10^{-7}\right)}$

Chapter 6

FACTORING AND APPLICATIONS

6.1   Factors; The Greatest Common Factor

Objective 1   Find the greatest common factor of a list of numbers.

Find the greatest common factor for each group of numbers.

1. 9, 12, 18   2. 36, 18, 24   3. 40, 25, 70

4. 108, 48, 84   5. 60, 75, 120   6. 17, 23, 40

7. 9, 18, 24, 48   8. 70, 126, 42, 56   9. 96, 480, 128

10. 84, 280, 112

Objective 2   Find the greatest common factor of a list of variables.

Find the greatest common factor for each list of terms.

11. $30x^3$, $40x^2$, $25x^2$

12. $18b^3$, $36b^6$, $45b^4$

13. $12ab^3$, $18a^2b^4$, $26ab^2$, $32a^2b^2$

14. $7m^4$, $12m^5$, $21m^9$

15. $24a^6$, $18a^7$, $42a^9$

16. y^7z^2, y^4z^8, z^3

17. $6k^2m^4n^5$, $8k^3m^7n^4$, k^4m^8,n^7

18. $29w^3x^7y^4$, $w^4x^5y^7$, $58w^2x^9,y^5$

19. $45a^7y^4$, $75a^3y^2$, $90a^2y$, $30a^4y^3$

20. $9xy^4,72x^4y^7$, $27xy^2$, $108x^2y^5$

Objective 3   Factor out the greatest common factor.

Complete the factoring.

21. $84 = 4(\quad)$

22. $18x^3 = 3x^2(\quad)$

23. $-18y^8 = -3y^5(\quad)$

24. $-75a^4y^2 = 25a^3y^2(\quad)$

Factor out the greatest common factor.

25. $18r + 24t$

26. $26r + 39t$

27. $18q^2 - 27q$

28. $45xy + 18x + 27x^3y$

29. $20x^2 + 40x^2y - 70xy^2$

30. $24ab - 8a^2 + 40ac$

31. $42tw + 21t - 63t^2$

32. $44rs - 22r + 77r^2$

33. $15a^7 - 25a^3 - 40a^4$

34. $9y^2 - 7$

35. $26x^8 - 13x^{12} + 52x^{10}$

36. $27r^2 - 54r^4 - 81r^5$

37. $56x^2y^4 - 24xy^3 + 32xy^2$

38. $3(a+b) - x(a+b)$

39. $9(x+8) + a(x+8)$

40. $x^2(r - 4s) + z^2(r - 4s)$

Objective 4   **Factor by grouping.**

Factor each polynomial by grouping.

41. $x^2 + 2x + 5x + 10$

42. $x^4 + 2x^2 + 5x^2 + 10$

43. $x^2 + 5x - 4x - 20$

44. $x^2 + 9x - 4x - 36$

45. $3x^2 - 9x + 12x - 36$

46. $8x^2 + 12xy - 2xy - 3y^2$

47. $xy - 2x - 2y + 4$

48. $15 - 5x - 3y + xy$

49. $5x + 15 - xy - 3y$

50. $6x^3 + 9x^2y^2 - 2xy^3 - 3y^5$

51. $2a^3 - 3a^2b + 2ab^2 - 3b^3$

52. $12x^3 - 4xy - 3x^2y^2 + y^3$

53. $2x^4 + 4x^2y^2 + 3x^2y + 6y^3$

54. $12x^2 + 4xy - 6xy - 2y^2$

6.1 Mixed Exercises

Find the greatest common factor for each list.

55. 56, 21, 49

56. 42, 48, 72

57. $15ab^2$, $45a^3b^4$, $70ab^3$

58. $51pq^3$, $34pq^2$, $68p^4q^5$

59. $35r^2s^5$, $25r^3t^2$, $28s^3t^4$

60. $-72u^2v^3$, $-54uv^2$, $-63uv^4$

Factor each polynomial completely.

61. $6xy^2 - 42xz^2$

62. $4pq^4 + 36q^2$

63. $45a^2b^3 - 90ab + 15ab^2$

64. $2a(x-2y) + 9b(x-2y)$

65. $m^2 - 5m + 3m - 15$

66. $2x^2 - 14xy + xy - 7y^2$

67. $15m^2 + 9m - 5m - 3$

68. $3r^3 - 2r^2s + 3s^2r - 2s^3$

69. $6r^2p + 4pq - 8p^2$

70. $1 + p - q - pq$

FACTORING AND APPLICATIONS

6.2 Factoring Trinomials

Objective 1 Factor trinomials with a coefficient of 1 for the squared term.

List all pairs of integers with the given product. Then find the pair whose sum is given.

1. Product: 42; sum: 17

2. Product: 28; sum: -11

3. Product: –64; sum: 12

4. Product: –54; sum –3

Complete the factoring.

5. $x^2 + 7x + 12 = (x+3)(\quad)$

6. $x^2 + 3x - 28 = (x-4)(\quad)$

7. $x^2 + 4x + 4 = (x+2)(\quad)$

8. $x^2 - x - 30 = (x+5)(\quad)$

Factor completely. If a polynomial cannot be factored, write **prime.**

9. $x^2 + 11x + 18$

10. $x^2 - 11x + 28$

11. $x^2 - x - 2$

12. $x^2 + 14x + 49$

13. $x^2 - 2x - 35$

14. $x^2 - 8x - 33$

15. $x^2 + 6x + 5$

16. $x^2 - 15xy + 56y^2$

17. $x^2 - 4xy - 21y^2$

18. $m^2 - 2mn - 3n^2$

Objective 2 Factor trinomials after factoring out the greatest common factor.

Factor completely.

19. $2x^2 + 10x - 28$

20. $3x^2 + 6x - 24$

21. $3h^3k - 21h^2k - 54hk$

22. $7b^2 - 42b + 56$

23. $4a^2 - 24b + 5$

24. $3p^6 + 18p^5 + 24p^4$

25. $2a^3b - 10a^2b^2 + 12ab^3$

26. $3y^3 + 9y^2 - 12y$

27. $5r^2 + 35r + 60$

28. $3xy^2 - 24xy + 36x$

29. $10k^6 + 70k^5 + 100k^4$

30. $x^5 - 3x^4 + 2x^3$

31. $2x^2y^2 + -2xy^3 - 12y^4$

32. $a^2b - 12ab^2 + 35b^3$

6.2 Mixed Exercises

Factor each polynomial completely. If a polynomial cannot be factored, write **prime**.

33. $a^2 - 10a + 21$

34. $p^2 - p - 12$

35. $x^3 - 4x^2 + 4x$

36. $5r^3 - 45r^2 + 70r$

37. $r^2 + r + 3$

38. $2m^3 - 2m^2 - 4m$

39. $2n^4 - 16n^3 + 30n^2$

40. $q^2 - 4q - 12$

41. $b^2 + 10bc + 25c^2$

42. $qr^3 - 4q^2r^2 - 21q^3r$

43. $a^2 + 10ab + 16b^2$

44. $3d^2 - 18d + 27$

45. $2s^2t + -6st - 20t$

46. $2x^3 - 14x^2y + 20xy^2$

FACTORING AND APPLICATIONS

6.3 Factoring Trinomials by Grouping

Objective 1 Factor trinomials by grouping when the coefficient of the squared term **is not 1**

Complete the factoring.

1. $2x^2 + 5x - 3 = (2x - 1)(\quad)$

2. $6x^2 + 19x + 10 = (3x + 2)(\quad)$

3. $16x^2 + 4x - 6 = (4x + 3)(\quad)$

4. $24y^2 - 17y + 3 = (3y - 1)(\quad)$

Factor each trinomial by grouping.

5. $8b^2 + 18b + 9$

6. $3x^2 + 13x + 14$

7. $15a^2 + 16a + 4$

8. $6n^2 + 11n + 4$

9. $3b^2 + 8b + 4$

10. $3m^2 - 5m - 12$

11. $3p^3 + 8p^2 + 4p$

12. $8m^2 + 26mn + 6n^2$

13. $7a^2b + 18ab + 8b$

14. $2s^2 + 5st - 3st^2$

15. $9c^2 + 24cd + 12d^2$

16. $25a^2 + 30ab + 9b^2$

6.4 Factoring Trinomials using FOIL

Objective 1 Factoring trinomials using FOIL.

17. $10x^2 + 19x + 6$

18. $4y^2 + 3y - 10$

19. $2a^2 + 13a + 6$

20. $5w^2 - 92 - 2$

21. $8q^2 + 10q + 3$

22. $8m^2 - 10m - 3$

23. $14b^2 + 3b - 2$

24. $15q^2 - 2q - 24$

25. $3a^2 + 8ab + 4b^2$

26. $9w^2 + 12wz + 4z^2$

27. $10c^2 - cd - 2d^2$

28. $6x^2 + xy - 12y^2$

29. $18x^2 - 27xy + 4y^2$

30. $12y^2 + 11y - 15$

6.3 and 6.4 Mixed Exercises

31. $3x^2 - 11x - 4$

32. $2p^2 + 11p + 5$

33. $6y^2 + y - 1$

34. $9y^2 - 16y - 4$

35. $3p^2 + 17p + 10$

36. $9r^2 + 12r - 5$

37. $7x^2 + 27x - 4$

38. $4c^2 + 14cd - 8d^2$

39. $2x^4 + 5x^3 - 12x^2$

40. $12a^3 + 26a^2b + 12ab^2$

41. $27r^2 + 6rt - 8t^2$

42. $2y^5z^2 - 5y^4z^3 - 3y^3z^4$

FACTORING AND APPLICATIONS

6.5 Special Factoring Techniques

Objective 1 Factor the difference of two squares.

Factor each binomial completely. If a binomial cannot be factored, write **prime**.

1. $x^2 + 16$

2. $x^2 - 49$

3. $100r^2 - 9s^2$

4. $y^2 - 64$

5. $25a^2 - 36$

6. $9j^2 - \frac{16}{49}$

7. $36 - 121d^2$

8. $121m^2 - 9n^2$

9. $x^4 - 81$

10. $9m^4 - 1$

11. $9x^4 - 16$

12. $81y^4 - 1$

13. $9x^2 + 16$

14. $m^4 n^2 - m^2$

Objective 2 Factor a perfect square trinomial.

Factor each trinomial completely. It may be necessary to factor out the greatest common factor first.

15. $y^2 + 6y + 9$

16. $q^2 + 14q + 49$

17. $m^2 - 8m + 16$

18. $c^2 + 22c + 121$

19. $z^2 - \frac{4}{3}z + \frac{4}{9}$

20. $4w^2 + 12w + 9$

21. $16q^2 - 40q + 25$

22. $9j^2 + 12j + 4$

23. $64p^4 + 48p^2q^2 + 9q^4$

24. $100p^2 - \frac{25}{2}pr + \frac{25}{64}r^2$

25. $9m^2 + .6m + .01$

26. $-16x^2 - 48x - 36$

27. $-12a^2 + 60ab - 75b^2$

28. $18x^2 + 84xy + 98y^2$

| Objective 3 | Factor a difference of cubes.

Factor.

29. $x^3 - y^3$

30. $8a^3 - 1$

31. $8r^3 - 27s^3$

32. $64x^3 - y^3$

33. $216m^3 - 125p^6$

34. $8a^3 - 125b^3$

35. $(r+s)^3 - 1$

36. $(m+n)^3 - (m-n)^3$

37. $x^3 - (x-1)^3$

38. $216x^3 - 8y^3$

| Objective 4 | Factor a sum of cubes.

Factor.

39. $x^3 + y^3$

40. $z^3 + 8$

41. $27r^3 + 8s^3$

42. $8a^3 + 64b^3$

43. $125p^3 + q^3$

44. $64x^3 + 343y^3$

45. $1 + (y+z)^3$

46. $(x-y)^3 + (x+y)^3$

47. $(a-1)^3 + a^3$

48. $t^3 + (t+2)^3$

6.5 Mixed Exercises

Factor each polynomial completely. If a polynomial cannot be factored, write **prime**.

49. $a^2 + 14a + 49$

50. $16a^2 - 25b^2$

51. $9p^2 - 121$

52. $4f^2 + 20f + 25$

53. $9x^2 + 49$

54. $49x^2 - 21x + \frac{9}{4}$

55. $a^4 - 8a^2b + 16b^2$

56. $z^2 - 1.6z + .64$

57. $j^2 + 8j + 16$

58. $\frac{1}{9}x^2 - 4xy + 36y^2$

59. $r^4 - 81x^4$

60. $-3x^2 - 24x - 48$

61. $-20y^2 - 20y - 5$

62. $12m^2 + 75$

63. $125m^3 + 8n^3$

64. $y^6 + 1$

65. $125m^3 - 64p^3$

66. $z^3 - 125y^3$

FACTORING AND APPLICATIONS

6.6 Solving Quadratic Equations by Factoring

Objective 1 Solve quadratic equations by factoring.

Solve each equation. Check your answers.

1. $(y+9)(2y-3)=0$

2. $(3k+4)(5k-7)=0$

3. $3x^2+7x+2=0$

4. $b^2-49=0$

5. $2x^2-3x-20=0$

6. $x^2-2x-63=0$

7. $8r^2=24r$

8. $3x^2-7x-6=0$

9. $3-5x=8x^2$

10. $9x^2+12x+4=0$

11. $25x^2=20x$

12. $9y^2=16$

13. $12x^2+7x-12=0$

14. $14x^2-17x-6=0$

15. $c(5c+17)=12$

16. $3x(x+3)=(x+2)^2-1$

Objective 2 Solve other equations by factoring.

Solve each equation.

17. $3x(x+7)(x-2)=0$

18. $x(2x^2-7x-15)=0$

19. $z(4z^2-9)=0$

20. $z^3-49z=0$

21. $25a=a^3$

22. $x^3+2x^2-8x=0$

23. $2m^3+m^2-6m=0$

24. $(4x^2-9)(x-2)=0$

25. $z^4+8z^3-9z^2=0$

26. $3z^3+z^2-4z=0$

27. $(x+4)(x^2+7x+10)=0$

28. $(y^2-5y+6)(y^2-36)=0$

29. $15x^2 = x^3 + 56x$

30. $(y-7)(2y^2 + 7y - 15) = 0$

31. $\left(x - \frac{3}{2}\right)(2x^2 + 11x + 15) = 0$

32. $(y-1)(y^2 - 25) = 0$

6.6 Mixed Exercises

Solve each equation.

33. $(3p-2)(p-7) = 0$

34. $x(x+5)(x-3) = 0$

35. $x^2 + 7x + 10 = 0$

36. $-9y = y^2$

37. $4x(x^2 + 4x + 4) = 0$

38. $3x^2 = 16x + 12$

39. $9p^2 - 25 = 0$

40. $49x^2 = x^4$

41. $(x-4)(x^2 - 9) = 0$

42. $3y^2 - 26y + 35 = 0$

43. $v^5 - 6v^4 + 8v^3 = 0$

44. $b(3b+16) = 0$

45. $-10x = -3x^2 + 8$

46. $(z-12)(z^3 - 4z) = 0$

FACTORING AND APPLICATIONS

6.7   Applications of Quadratic Equations

Objective 1   **Solve problems about geometric figures.**

Solve each equation.

1. The length of a rectangle is 8 centimeters more than the width. The area is 153 square centimeters. Find the length and width of the rectangle.

2. The length of a rectangle is three times its width. If the width were increased by 4 and the length remained the same, the resulting rectangle would have an area of 231 square inches. Find the dimensions of the original rectangle.

3. The area of a rectangular room is 252 square feet. Its width is 4 feet less than its length. Find the length and width of the room.

4. Two rectangles with different dimensions have the same area. The length of the first rectangle is three times its width. The length of the second rectangle is 4 meters more than the width of the first rectangle, and its width is 2 meters more than the width of the first rectangle. Find the lengths and widths of the two rectangles.

5. Each side of one square is 1 meter less than twice the length of each side of a second square. If the difference between the areas of the two squares is 16 meters, find the lengths of the sides of the two rectangles.

6. The area of a triangle is 42 square centimeters. The base is 2 centimeters less than twice the height. Find the base and height of the triangle.

7. A rectangular bookmark is 6 centimeters longer than it is wide. Its area is numerically 3 more than its perimeter. Find the length and width of the bookmark.

8. A book is three times as long as it is wide. Find the length and width of the book in inches if its area is numerically 128 more than its perimeter.

9. The volume of a box is 192 cubic feet. If the length of the box is 8 feet and the width is 2 feet more than the height, find the width of the box.

10. Mr. Fixxall is building a box which will have a volume of 60 cubic meters. The height of the box will be 4 meters, and the length will be 2 meters more than the width. Find the width of the box.

Objective 2   **Solve applied problems about consecutive integers.**

Solve each problem.

11. The product of two consecutive integers is four less than four times their sum. Find the integers.

12. Find two consecutive integers such that the square of their sum is 169.

13. The product of two consecutive positive odd integers is 1 less than four times their sum. Find the integers.

14. Find two consecutive integers such that the sum of the squares of the two integers is 3 more than the opposite (additive inverse) of the smaller integer.

15. If the square of the sum of two consecutive integers is reduced by twice their product, the result is 25. Find the integers.

16. The product of two consecutive even integers is 24 more than three times the larger integer. Find the integers.

17. Find all possible pairs of consecutive odd integers whose sum is equal to their product decreased by 47.

18. Find two consecutive positive even integers whose product is 168.

19. Find two consecutive positive even integers whose product is 6 more than three times its sum.

20. Find two consecutive integers whose product is three more than three times its sum.

Objective 3   **Solve problems using the Pythagorean formula.**

Solve each problem.

21. The hypotenuse of a right triangle is 4 inches longer than the shorter leg. The longer leg is 4 inches shorter than twice the shorter leg. Find the length of the shorter leg.

22. A flag is shaped like a right triangle. The hypotenuse is 6 meters longer than twice the length of the shortest side of the flag. If the length of the other side is 2 meters less than the hypotenuse, find the lengths of the sides of the flag.

23. A field has a shape of a right triangle with one leg 10 meters longer than twice the length of the other leg. The hypotenuse is 4 meters longer than three times the length of the shorter leg. Find the dimensions of the field.

24. A train and a car leave a station at the same time, the train traveling due north and the car traveling west. When they are 100 miles apart, the train has traveled 20 miles farther than the car. Find the distance each has traveled.

25. The hypotenuse of a right triangle is 1 foot larger than twice the shorter leg. The longer leg is 7 feet larger than the shorter leg. Find the length of the longer leg.

26. Mark is standing directly beneath a kite attached to a string which Nina is holding, with her hand touching the ground. The height of the kite at that instant is 12 feet less than twice the distance between Mark and Nina. The length of the kite string is 12 feet more than that distance. Find the length of the kite string.

27. A 30-foot ladder is leaning against a building. The distance from the bottom of the ladder to the building is 6 feet less than the distance from the top of the ladder to the ground. How far is the bottom of the ladder from the building?

28. A field is in the shape of a right triangle. The shorter leg measures 45 meters. The hypotenuse measures 45 meters less than twice the longer the leg. Find the dimensions of the lot.

29. Two ships left a dock at the same time. When they were 25 miles apart, the ship that sailed due south had gone 10 miles less than twice the distance traveled by the ship that sailed due west. Find the distance traveled by the ship that sailed due south.

30. A ladder is leaning against a building. The distance from the bottom of the ladder to the building is 8 feet less than the length of the ladder. How high up the side of the building is the top of the ladder if that distance is 4 feet less than the length of the ladder?

Objective 4   Solving problems using given quadratic models.

6.7 Mixed Exercises

31. The length of a rectangular picture is 4 centimeters more than three times the width. The area is 64 square centimeters. Find the length and width of the picture.

32. The area of a rectangular garden plot is 72 square feet. If the length of the plot is 6 feet less than three times the width, find the dimensions of the garden plot.

33. The product of two consecutive even positive integers is 10 more than seven times the larger. Find the integers.

34. Find three consecutive positive odd integers such that four times the sum of all three equals 13 more than the product of the smaller two.

35. The length of the shorter leg of a right triangle is tripled and 2 inches is subtracted from the result, giving the length of the hypotenuse. The longer leg is 2 inches longer than twice the shorter leg. Find the length of the shorter leg of the triangle.

36. Michael wishes to build a box to hold his tools. The box is to be 4 feet high, and the width of the box is to be 3 foot less than the length. The volume of the box will be 160 cubic feet. Find the length and width of the box.

37. If the sum of the squares of two consecutive positive integers is reduced by their product, the result is 21. Find the integers.

38. Two trains leave New York City at the same time. One train travels due north and the other travels due east. When they are 75 miles apart, the train going north has gone 30 miles less than twice the distance traveled by the train going east. Find the distance traveled by the train going east.

39. Penny and Carla started walking from the same corner. Penny walked east and Carla walked south. When they were 26 miles apart, Carla had walked 14 miles further than Penny. Find the distance each had walked.

40. The length of a rectangular label is 2 centimeters more than three times its width. If the length was decreased by 1 while the width stayed the same, the area of the new label would be 80 square centimeters. Find the length and width of the original label.

41. Linda has a rectangular piece of carpet whose length is 5 yards shorter than twice its width. The area of the piece is 88 square yards. Find the width of the piece.

42. Dipti and Yohannes left Times Square at the same time. Dipti traveled east and Yohannes traveled south. When Dipti traveled 14 miles farther than Yohannes, the distance between them was 2 miles greater than the distance Dipti traveled. What was the distance between them at that time?

43. The product of the smaller two of three consecutive positive odd integers equals 1 less than four times the largest. Find the integers.

44. The sides of one square have a length 2 meters more than the sides of another square. If the area of the larger square is subtracted from twice the area of the smaller square, the answer is 8 square meters. Find the lengths of the sides of each square.

45. If an object is propelled upward from a height of 16 feet with an initial velocity of 48 feet per second, its height h (in feet) t seconds later is given by the equation

$$h = -16t^2 + 48t + 16$$

(a)   After how many seconds is the height 52 feet?

(b)   After how many seconds is the height 48 feet?

46. A ball is dropped from the roof of a 19.6 meter high building. Its height h (in meters) t seconds later is given by the equation

$$h = -4.9t^2 + 19.6$$

(a)   After how many seconds is the height 18.375 meters?

(b)   After how many seconds is the height 14.7 meters?

(c)   After how many seconds does the ball hit the ground?

Chapter 7

RATIONAL EXPRESSIONS AND APPLICATIONS

7.1   Rational Expressions and Functions; Multiplying and Dividing

Objective 1   Define rational expressions.

Objective 2   Define rational functions and describe their domains.

Find all numbers that are not in the domain of the function.

1. $f(x) = \dfrac{4}{x-5}$

2. $f(x) = \dfrac{x}{x+5}$

3. $f(x) = \dfrac{2x+3}{x+7}$

4. $f(m) = \dfrac{m+7}{m}$

5. $f(t) = \dfrac{t-9}{4}$

6. $f(a) = \dfrac{2a-3}{4a-7}$

7. $f(s) = \dfrac{8s+7}{3s-2}$

8. $f(r) = \dfrac{r+7}{r^2-25}$

9. $f(q) = \dfrac{q+7}{q^2-3q+2}$

10. $f(x) = \dfrac{2x-5}{x^2-10x+25}$

11. $f(x) = \dfrac{x-6}{x^2+1}$

12. $f(x) = \dfrac{x^2-4}{x^2+4}$

Objective 3   Write rational expressions in lowest terms.

Write in lowest terms.

13. $\dfrac{4n^5}{16n^3}$

14. $\dfrac{35p^5}{15}$

15. $\dfrac{-6x^3y^7}{18xy^4}$

16. $\dfrac{(m-7)(m+3)}{(m+3)(m-2)}$

17. $\dfrac{k(k+4)}{5k(k+4)}$

18. $\dfrac{3x-3}{4x-4}$

19. $\dfrac{6y^2+y}{3y^2+y}$

20. $\dfrac{x^2-6x+9}{x^2-9}$

21. $\dfrac{11r^2-22r^3}{6-12r}$

22. $\dfrac{s^2-s-6}{s^2+s-12}$

23. $\dfrac{8z^2+6z-9}{16z^2-9}$

24. $\dfrac{-x+y}{y-x}$

25. $\dfrac{c-2d}{2d-c}$

26. $\dfrac{x^2-1}{1-x}$

27. $\dfrac{a^2-3a}{3a-a^2}$

28. $\dfrac{r^3-s^3}{r^2-s^2}$

29. $\dfrac{x^2-y^2}{x^3-y^3}$

30. $\dfrac{(p-5)(5-p)}{(5-p)(5+p)}$

Objective 4   **Multiply rational expressions.**

Multiply.

31. $\dfrac{9x^2}{16} \cdot \dfrac{4}{3x}$

32. $\dfrac{21z^4}{8z} \cdot \dfrac{4z^3}{7z^5}$

33. $\dfrac{4r^2}{8r} \cdot \dfrac{3r^3}{4r^4}$

34. $\dfrac{6y^3}{9y} \cdot \dfrac{12y}{y^2}$

35. $\dfrac{10a^3}{20a^2} \cdot \dfrac{12a^4}{3a}$

36. $\dfrac{m+3}{2} \cdot \dfrac{12}{(m+3)^2}$

37. $\dfrac{3(s-1)}{s} \cdot \dfrac{2s}{5(s-1)}$

38. $\dfrac{5t+25}{10} \cdot \dfrac{12}{6t+30}$

39. $\dfrac{9k-18}{6k+12} \cdot \dfrac{3k+6}{15k-30}$

40. $\dfrac{x^2-6x}{9x} \cdot \dfrac{18x}{3x-18}$

41. $\dfrac{x^2-x-6}{x^2-2x-8} \cdot \dfrac{x^2+7x+12}{x^2-9}$

42. $\dfrac{6z^2-5z-6}{6z^2+5z-6} \cdot \dfrac{12z^2-17z+6}{12z^2-z-6}$

Objective 5   **Find reciprocals for rational expressions.**

Find the reciprocal.

43. $\dfrac{3}{y}$

44. $\dfrac{4}{p-5}$

45. $\dfrac{x^2+9}{7x}$

46. $\dfrac{m^2+2m+3}{5}$

47. $\dfrac{2p-1}{p^2+7p}$

48. $\dfrac{n+8}{7}$

49. $\dfrac{8-s}{s-8}$

50. $\dfrac{r^2+2r}{5+r}$

51. $\dfrac{x^2+4}{3x-6}$

52. $\dfrac{7z+7}{z^2-9}$

53. 0

54. $\dfrac{x^2-3x+4}{x^2+x+2}$

Objective 6   **Divide rational expressions.**

Divide.

55. $\dfrac{7x^4}{6x^2} \div \dfrac{14x^3}{3x}$

56. $\dfrac{15t^{10}}{9t^5} \div \dfrac{6t^6}{10t^4}$

57. $\dfrac{3s^4}{4s^3} \div \dfrac{9s^3}{32s^4}$

58. $\dfrac{5r^3}{4r^2} \div \dfrac{15r^2}{8r^4}$

59. $\dfrac{k-3}{16} \div \dfrac{k-3}{32}$

60. $\dfrac{12}{4x-12} \div \dfrac{2}{x-3}$

61. $\dfrac{2n+8}{6} \div \dfrac{3n+12}{2}$

62. $\dfrac{12p+24}{36p-36} \div \dfrac{6p+12}{8p-8}$

63. $\dfrac{a^2-16}{a+3} \div \dfrac{a-4}{a^2-9}$

64. $\dfrac{6(m+2)}{3(m-1)^2} \div \dfrac{(m+2)^2}{9(m-1)}$

65. $\dfrac{4z+12}{2z-10} \div \dfrac{z^2-9}{z^2-z-20}$

66. $\dfrac{8-y}{y-8} \div \dfrac{y-8}{y+8}$

67. $\dfrac{x^2-x-6}{x^2+x-12} \div \dfrac{x^2+2x-3}{x^2+3x-4}$

68. $\dfrac{16-r^2}{r^2+2r-8} \div \dfrac{r^2-2r-8}{4-r^2}$

69. $\dfrac{s^2-s-2}{s^2+5s+4} \div \dfrac{s-2}{s+3}$

70. $\dfrac{2a^2-5a-12}{a^2-10a+24} \div \dfrac{4a^2-9}{a^2-9a+18}$

7.1   Mixed Exercises

Write each rational expression in lowest terms, and multiply or divide as indicated.

71. $\dfrac{6q^2r}{30qr^3}$

72. $\dfrac{6-3x}{2x-4}$

73. $\dfrac{z^2-9z}{9-z}$

74. $\dfrac{8r+r^2}{r+8}$

75. $\dfrac{2-y}{8} \cdot \dfrac{7}{y-2}$

76. $\dfrac{5-x}{4} \cdot \dfrac{12}{x-5}$

77. $\dfrac{(t+1)^3(t+4)}{t^2+5t+4} \div \dfrac{t^2+2t+1}{t^2+3t+2}$

78. $\dfrac{x^2-25}{5+x} \cdot \dfrac{x}{5-x}$

79. $\dfrac{p^2+3p+2}{p^2-3p-4} \cdot \dfrac{p^2-10p+24}{p^2+5p+6}$

80. $\dfrac{(z+4)^2(z-1)}{z^2+3z-4} \div \dfrac{z^2-16z}{z^2+8z+16}$

7.2   Adding and Subtracting Rational Expressions

Objective 1   **Add and subtract rational expressions with the same denominator.**

Add or subtract as indicated. Write all answers in lowest terms.

1. $\dfrac{3}{x} + \dfrac{8}{x}$

2. $\dfrac{5}{y^2} - \dfrac{8}{y^2}$

3. $\dfrac{3}{5t} + \dfrac{15}{5t}$

4. $\dfrac{c}{5a} - \dfrac{4}{5a}$

5. $\dfrac{n}{m+3} - \dfrac{-3n+7}{m+3}$

6. $\dfrac{z^2}{z-y} - \dfrac{y^2}{z-y}$

7. $\dfrac{r}{r^2-s^2} + \dfrac{s}{r^2-s^2}$

8. $\dfrac{4x+3}{x-7} - \dfrac{3x+10}{x-7}$

9. $\dfrac{x}{x^2-7x+10} - \dfrac{2}{x^2-7x+10}$

10. $\dfrac{k}{k^2+3k-10} - \dfrac{2}{k^2+3k-10}$

11. $\dfrac{1}{q^2-6q-7} + \dfrac{q}{q^2-6q-7}$

12. $\dfrac{b}{a^2-b^2} - \dfrac{a}{a^2-b^2}$

Objective 2   **Find a least common denominator.**

Assume that the expressions given are denominators of fractions. Find the least common denominator (LCD) for each group.

13. $5m, 6m$

14. $25z, 30z$

15. $5x, 15x^2, 25xy$

16. $t, t-1$

17. $8s + 24, 3s + 9$

18. $5a + 10, a^2 + 2a$

19. $q^2 - 36, (q+6)^2$

20. $r - p, p - r$

21. $r^2 + 5r + 4, r^2 + r$

22. $3n + n^2, 3 - n$

23. $p - 4, p^2 - 16, (p+4)^2$

24. $2z^2 + 7z - 4, 2z^2 - 7z + 3$

Objective 3   **Add and subtract rational expressions with different denominators.**

Add or subtract as indicated. Write all answers in lowest terms.

25. $\dfrac{5}{y} + \dfrac{4}{7}$

26. $\dfrac{9}{x} + \dfrac{3}{2}$

27. $\dfrac{3}{5} - \dfrac{1}{z}$

28. $\dfrac{5a}{6} - \left(\dfrac{2a}{3} - \dfrac{a}{6} \right)$

29. $\dfrac{4+2m}{5}+\dfrac{2+m}{10}$

30. $\dfrac{6}{s^2}-\dfrac{2}{s}$

31. $\dfrac{6}{5t}+\dfrac{4}{3t^2}$

32. $\dfrac{3r+4}{3}+\dfrac{6r+4}{6}$

33. $\dfrac{1}{x^2-1}-\dfrac{1}{x^2+3x+2}$

34. $\dfrac{y+9}{y^2-16}+\dfrac{2}{y+4}$

35. $\dfrac{4}{2-a}+\dfrac{7}{a-2}$

36. $\dfrac{-1}{3-m}-\dfrac{2}{m-3}$

37. $\dfrac{5r}{r+2s}-\dfrac{3r}{-r-2s}$

38. $\dfrac{a+3b}{b^2+2ab+a^2}+\dfrac{a-b}{3b^2+4ab+a^2}$

7.2   Mixed Exercises

Find the least common denominator for each group of rational expressions.

39. $\dfrac{-7}{2-m},\ \dfrac{3}{m-2}$

40. $\dfrac{x+5}{x^2+5x+6},\ \dfrac{3-x}{3x+6}$

Add or subtract as indicated. Write all answers in lowest terms.

41. $\dfrac{2x-1}{x^2+x-2}-\dfrac{x}{x^2+x-2}$

42. $\dfrac{1}{r-6}-\dfrac{2}{6-r}$

43. $\dfrac{4}{3a}+\dfrac{7}{2a^2b}$

44. $\dfrac{q+2}{q}+\dfrac{q}{q+2}$

45. $\dfrac{8}{k-2}-\dfrac{4}{k+2}$

46. $\dfrac{6}{m+n}+\dfrac{2}{m-n}$

47. $\dfrac{z}{z^2-1}+\dfrac{z-1}{z^2+2z+1}$

48. $\dfrac{2}{4y^2-16}+\dfrac{3}{4+2y}$

7.3   Complex Fractions

Objective 1   Simplify complex fractions by simplifying the numerator and denominator. (Method 1)

Use Method 1 to simplify.

1. $\dfrac{\frac{x+1}{x}}{\frac{x+1}{y}}$

2. $\dfrac{\frac{2p+3}{5}}{\frac{8p+12}{3}}$

3. $\dfrac{\frac{3}{k}+1}{\frac{3+k}{2}}$

4. $\dfrac{\frac{1}{t}+\frac{1}{z}}{\frac{1}{z+t}}$

5. $\dfrac{\frac{1}{s}+r}{\frac{1}{r}+s}$

6. $\dfrac{\frac{m+1}{m-1}}{\frac{1}{m+1}}$

7. $\dfrac{q+\frac{1}{q+1}}{q-\frac{1}{q}}$

8. $\dfrac{\frac{1}{a+b}}{\frac{4}{a^2-b^2}}$

9. $\dfrac{\frac{rs}{3r^2}}{\frac{s^2}{3}}$

Objective 2   Simplify complex fractions by multiplying by a common denominator. (Method 2)

Use Method 2 to simplify.

10. $\dfrac{\frac{3}{k}}{\frac{9}{k^2}}$

11. $\dfrac{\frac{6}{m}}{\frac{12}{m^2}}$

12. $\dfrac{x+\frac{2}{x}}{\frac{x^2+2}{3}}$

13. $\dfrac{\frac{a-2b}{a}}{\frac{a-2b}{b}}$

14. $\dfrac{\frac{s}{s+1}}{\frac{5}{2(s+1)}}$

15. $\dfrac{p+\frac{1}{p}}{\frac{3}{p}-p}$

16. $\dfrac{t-\frac{2}{t}}{t+\frac{4}{t}}$

17. $\dfrac{\frac{2}{x-1}+2}{\frac{2}{x+1}-2}$

18. $\dfrac{\frac{r}{r+1}+1}{\frac{2r+1}{r-1}}$

Objective 3   Compare the two methods of simplifying complex fractions.

Use either method to simplify each complex fraction.

19. $\dfrac{\frac{m+n}{m}}{\frac{1}{m}+\frac{1}{n}}$

20. $\dfrac{a+\frac{1}{a}}{\frac{1}{a}-a}$

21. $\dfrac{\frac{z-y}{z+y}}{\frac{z}{z-y}}$

22. $\dfrac{\frac{1}{t+1}-1}{\frac{1}{t-1}+1}$

23. $\dfrac{\frac{25k^2-m^2}{4k}}{\frac{5k+m}{7k}}$

24. $\dfrac{\frac{1}{r}}{\frac{1+r}{1-r}}$

Objective 4   Simplify rational expressions with negative exponents.

Simplify each expression, using only positive exponents in the answer.

25. $\dfrac{x^{-1}}{x^{-1}+5}$

26. $\dfrac{1+x^{-1}}{(x+y)^{-1}}$

27. $x^{-1}+y^{-2}$

28. $\dfrac{2x^{-1}+y^2}{z^{-3}}$

29. $\dfrac{z^{-2}}{4+y^{-3}}$

31. $\dfrac{4x^{-2}}{2+6y^{-3}}$

33. $\dfrac{x^{-1}}{y-x^{-1}}$

30. $(x^{-2}-y^{-2})^{-1}$

32. $\dfrac{s^{-1}+r}{r^{-1}+s}$

34. $\dfrac{(m+n)^{-2}}{m^{-2}-n^{-2}}$

7.3   Mixed Exercises

Use either method to simplify each complex fraction.

35. $\dfrac{\frac{1}{x}+\frac{1}{x-1}}{\frac{1}{x}-\frac{2}{x-1}}$

37. $\dfrac{\frac{m}{n+1}}{\frac{m^2}{n}}$

39. $\dfrac{\frac{m}{4}-\frac{1}{m}}{1+\frac{m+4}{m}}$

36. $\dfrac{\frac{5p}{y^3}}{\frac{5p^2}{y^3}}$

38. $\dfrac{n-\frac{n+2}{4}}{\frac{3}{4}-\frac{5}{2n}}$

40. $\dfrac{\frac{s}{s+1}}{\frac{2}{s^2-1}}$

Simplify each expression, using only positive exponents in the answer.

41. $\dfrac{x^{-1}-y^{-1}}{x^{-1}+y^{-1}}$

42. $(x^{-4}+4)^{-1}$

43. $\dfrac{x^{-1}-x}{x^{-1}+1}$

44. $\dfrac{x^{-2}-y^{-2}}{x+y}$

7.4   Equations with Rational Expressions and Graphs

Objective 1   Determine the domain of a rational equation.

(a) Without actually solving the equations below, list all possible numbers that would have to be rejected if they appeared as potential solutions. (b) Then give the domain using set notation.

1. $\dfrac{1}{x} + \dfrac{2}{x+1} = 0$

2. $\dfrac{4}{2x-5} - \dfrac{6}{3x+1} = \dfrac{1}{2}$

3. $\dfrac{10}{x-7} + \dfrac{7}{8+x} = 0$

4. $\dfrac{x}{6} - \dfrac{1}{2} = \dfrac{3}{x+1}$

5. $\dfrac{4}{x^2-2x} + \dfrac{x}{2} = 0$

6. $\dfrac{3}{3+x} + \dfrac{5}{x^2-x} = \dfrac{9}{x}$

Objective 2   Solve rational equations.

Solve each equation.

7. $\dfrac{6}{x} = 4 - \dfrac{5}{x}$

8. $\dfrac{2p}{7} - 5 = p$

9. $\dfrac{z+1}{2} = \dfrac{z+2}{3}$

10. $\dfrac{9}{m-2} = 3$

11. $\dfrac{2y+3}{y} = \dfrac{3}{2}$

12. $\dfrac{5-a}{a} + \dfrac{3}{4} = \dfrac{7}{a}$

13. $\dfrac{5x-3}{4x+7} = \dfrac{7}{15}$

14. $\dfrac{2}{t} = \dfrac{t}{5t-12}$

15. $\dfrac{x}{2x+2} = \dfrac{-2x}{4x+4} + \dfrac{2x-3}{x+1}$

16. $\dfrac{4}{n} - \dfrac{2}{n+1} = 3$

Objective 3 **Recognize the graph of a rational function.**

Graph each rational function. Give the equation of the vertical asymptote.

17. $f(x) = \dfrac{5}{x}$

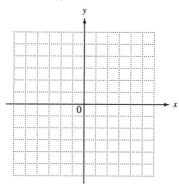

19. $f(x) = \dfrac{2}{x-1}$

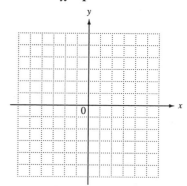

18. $f(x) = \dfrac{1}{x-3}$

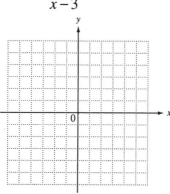

20. $f(x) = \dfrac{1}{x-4}$

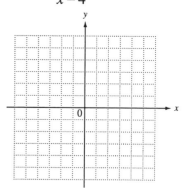

7.4   Mixed Exercises

Solve each equation.

21. $\dfrac{2w+1}{5} = \dfrac{w+1}{4}$

22. $\dfrac{x}{8} + \dfrac{x}{4} = 8$

23. $\dfrac{3}{4p} - \dfrac{2}{p} = \dfrac{5}{12}$

24. $\dfrac{-3}{r} = 2 + \dfrac{1}{r}$

25. $\dfrac{x+3}{x} - \dfrac{x+4}{x+5} = \dfrac{15}{x^2+5x}$

26. $\dfrac{y}{y-4} + \dfrac{2}{y} = \dfrac{16}{y^2-4y}$

27. $\dfrac{q+12}{q^2-16} - \dfrac{3}{q-4} = \dfrac{1}{q+4}$

28. $x - 11 + \dfrac{18}{x} = 0$

29. $\dfrac{3p}{p^2+5p+6} = \dfrac{5p}{p^2+2p-3} - \dfrac{2}{p^2+p-2}$

30. $\dfrac{1}{b+2} - \dfrac{5}{b^2+9b+14} = \dfrac{-3}{b+7}$

7.5   Applications of Rational Expressions

Objective 1   Find the value of an unknown variable in a formula.

Find the value of the variable indicated.

1. If $m = \frac{a}{b}$, $m = 9$, and $a = 5$, find b.

2. If $I = \frac{E}{R}$, $I = 12$, and $E = 4$, find R.

3. If $F = \frac{GmM}{d^2}$, $F = 150$, $G = 32$, $M = 50$, and $d = 10$, find m.

4. If $c = \frac{100b}{L}$, $c = 80$, and $b = 16$, find L.

5. If $\frac{1}{R} = \frac{1}{R_1} + \frac{1}{R_2}$, $R = 10$, and $R_1 = 20$, find R_2.

6. If $\frac{1}{f} = \frac{1}{d_0} + \frac{1}{d_i}$, $f = 10$, and $d_0 = 25$, find d_i.

7. If $r = \frac{d}{t}$, $r = 60$, and $d = 120$, find t.

8. If $t = \frac{I}{pr}$, $t = 3$, $p = 100$, and $I = 15$, find r.

9. If $h = \frac{2A}{B+b}$, $A = 40$, $h = 8$, and $b = 3$, find B.

10. If $B = \frac{3V}{h}$, $B = 20$, and $V = 80$, find h.

Objective 2   Solve a formula for a specified variable.

Solve each formula for the specified variable.

11. $\dfrac{1}{R} = \dfrac{1}{R_1} + \dfrac{1}{R_2}$ for R_2

12. $\dfrac{1}{f} = \dfrac{1}{d_0} + \dfrac{1}{d_i}$ for d_0

13. $\dfrac{V_1 P_1}{T_1} = \dfrac{V_2 P_2}{T_2}$ for T_2

14. $F = \dfrac{Gm_1 m_2}{d^2}$ for m_2

15. $s_n = \dfrac{n}{2}(a_1 + a_n)$ for a_n

16. $A = \dfrac{1}{2} h(b_1 + b_2)$ for b_1

17. $\dfrac{V_1 P_1}{T_1} = \dfrac{V_2 P_2}{T_2}$ for V_1

18. $A = \dfrac{R_1 R_2}{R_1 + R_2}$ for R_1

19. $I = \dfrac{nE}{R + nr}$ for R

20. $E = \dfrac{e(R + r)}{r}$ for r

Objective 3 **Solve applications using proportions.**

Use proportions to solve each problem.

21. A student is expected to answer 5 of 6 questions correctly on a certain pretest. If there are 24 questions on the test, how many questions is a student expected to answer correctly?

22. In a certain midwestern city in a recent year, there were 500 crimes committed per 100,000 population. If the population of that city was 350,000, how many crimes were committed?

23. If Jennifer can address her wedding invitations in $4\frac{1}{2}$ hours, what is her rate (in job per hour)?

24. Alex paid $1.24 in sales tax on a purchase of $15.50. How much sales tax would he pay on a purchase of $480 at the same tax rate?

25. Steven invested $20,000 and earned $900 in income in the first year. How much income would he have earned if he had invested $25,000 in the same account?

26. Holly drove 315 miles in 7 hours. How long would it take her to drive 225 miles if she traveled at the same rate?

27. Ryan's car travels 24 miles using 1 gallon of gas. If Ryan has 4 gallons of gas in his car, how much more gas will he need to travel 288 miles?

28. Connie paid $.90 in sales tax on a purchase of $12.00. Later that day she purchased an item for which she paid $55.90 including sales tax. How much was the sales tax on that item?

29. A certain city with a population of 400,000 had 1600 burglaries during the last year. If the number of burglaries increased by 200 this year and its burglary rate remained the same, by what number did its population increase?

30. Megan has invested $5000 in an account that earns $200 in income per year. If she wishes to earn $240 per year at the same rate, by how much must she increase her investment.

Objective 4 **Solve applications about distance, rate, and time.**

Solve each problem.

31. Kate averages 10 miles per hour riding her bike to town. Averaging 30 miles per hour by car takes her 2 hours less time. How far does she travel to town?

32. Emily's boat goes 14 miles per hour. Find the speed of the current in the river if she can go 8 miles downstream in the same time as she can go 6 miles upstream.

33. Lauren's boat can go 9 miles per hour in still water. How far downstream can Lauren go if the river has a current of 3 miles per hour and she must be back in 4 hours?

34. A canal has a current of 2 miles per hour. Find the speed of Leroy's boat in still water if it goes 30 miles downstream in the same time as it takes to go 18 miles upstream.

35. Wilmoth can fly his plane 180 miles against the wind in the same time it takes him to fly 540 miles with the wind. The wind blows at 30 miles per hour. Find the speed of his plane in still air.

36. Pauline and Pete agree to meet in Columbia. Pauline travels 120 miles, while Pete travels 80 miles. If Pauline's speed is 20 miles per hour greater than Pete's and they both spend the same amount of time traveling, at what speed does each travel?

37. Tim traveled from here to Bristol at 20 miles per hour and from Bristol to here at 60 miles per hour. If the total time was 4 hours, how far is it from here to Bristol?

38. A river has a current of 3 kilometers per hour. Find the speed of Jena's boat in still water if she travels 60 kilometers downstream in the same time as 36 kilometers upstream.

39. Leo can get to the lake using either the old road at 40 miles per hour, or the new road at 60 miles per hour. If both roads are the same length and he gets there 1 hour sooner on the new road, how far is it to the lake?

40. Olivia can ride her bike 4 miles per hour faster than Ted. If Olivia can go 30 miles in the same time that Ted can go 15 miles, what are their speeds?

Objective 5   **Solve applications about work rates.**

Solve each problem.

41. Rose and Mose want to paint a chair. Mose can do it in 3 hours while Rose can do it in 4 hours. How long will it take them working together?

42. Rod can do a job in 3 hours, while Karen requires 9 hours. How long will it take them if they work together?

43. Amy and Anne (identical twins) are cleaning up their room. Amy can do it in 8 hours, but Anne needs 11 hours. How long will it take them if they work together?

44. Jason can erase the blackboard in 11 minutes while Rebecca can do it in 13 minutes. How long will it take them if they work together?

45. A perfume factory has a vat for perfume. An inlet pipe can fill it in 10 hours, while an outlet pipe can empty it in 12 hours. How long will it take to fill the vat if both the outlet and inlet pipes are left open?

46. A vat of chocolate can be filled by an inlet pipe in 12 hours. An outlet pipe can empty the vat in 15 hours. How long will it take to fill the vat if both pipes are left open?

47. An inlet pipe can fill a barrel with wine in 8 hours, while an outlet pipe can empty it in 6 hours. Through an error both pipes are left open. How long will it take to empty the barrel?

48. Glenn can weed the garden in 3 hours while Mike can weed it in 5 hours. How long will it take them if they work together?

49. Michelle can type a report in 4 hours but Beth can do it in 1 hour. How long will it take them if they work together?

50. Sara can mow the lawn in 5 hours and Casey can do it in 6 hours. How long will it take if they work together?

7.5   Mixed Exercises

Find the value of the variable indicated.

51. If $F = \frac{9}{5}C + 32$, and $F = 98.6$, find C.

52. If $A = \frac{1}{2}h(B+b)$, $B = 18$, $h = 4$, and $A = 60$, find b.

53. If $\frac{1}{x} = \frac{1}{y} + \frac{1}{z}$, $x = 5$, and $y = 3$, find z.

54. If $\frac{mn}{p} = \frac{qr}{s}$, $m = 4$, $n = 2.5$, $q = 7.5$, $r = 2.5$, and $s = 1.5$, find p.

Solve each formula for the specified variable.

55. $C = \frac{5}{9}(F - 32)$ for F

57. $\frac{1}{p} + \frac{1}{8} = \frac{7}{r}$ for r

59. $F = f\left[\dfrac{v + v_0}{v - v_s}\right]$ for v_s

56. $\frac{6}{5x} - \frac{7}{8y} = \frac{3}{z}$ for x

58. $h = \dfrac{2A}{B + b}$ for B

60. $F = f\left[\dfrac{v + v_0}{v - v_s}\right]$ for v_0

Solve each problem.

61. Vince can clean the office he shares with Miklos in 5 hours, while Miklos can do the job in 4 hours. How long will it take them if they work together?

62. Mark's boat goes 10 miles per hour. Find the speed of the current of the river if he can go 30 miles upstream in the same time that he takes to go 70 miles downstream.

63. In basketball, Wendi hopes to complete 4 baskets for every 7 baskets she attempts. If she attempts 35 baskets, how many should she complete to fulfill her hopes?

64. If a vat of tomato juice at a cannery can be filled by an inlet pipe in 12 hours, and emptied by an outlet pipe in 18 hours, how long will it take to fill the vat if both pipes must remain open?

65. A stream has a current of 2 miles per hour. Find the speed of Barbara's boat in still water if it goes 16 miles downstream in the same time it takes to go 12 miles upstream.

66. Mike paid $5.12 sales tax on an item priced $64. Later he made a purchase that cost him a total of $51.84, including sales tax at the same rate. How much sales tax did he pay on the second item?

67. At 8:00 A.M. Richard left home to walk 5 miles to a friend's house. After visiting his friend for 1 hour, Richard's wife picked him up and drove home at a speed of 30 miles per hour. They arrived home at 10:25 A.M. What was Richard's speed walking to his friend's house?

68. Vivian has invested $4000 in an account that earned $180 in the past year. She reinvested the amount she earned in the same account without withdrawing any of the original amount. If the account earns income at the same rate during the next year, how much income will Vivian earn next year?

69. Lyle can fly her plane 250 miles against the wind in the same time that it takes her to fly 300 miles with the wind. If the wind blows at 25 miles per hour, find the speed of Lyle's plane in still air.

70. Suppose that Joe and Marty can clean their entire house in 6 hours, while their toddler, Midgie, just by being around, can completely mess it up in $1\frac{1}{2}$ hours. If Midgie comes home to a clean house after visiting her grandma and Joe and Marty start to clean up the minute she gets home, how long does it take until the house is in shambles?

7.6   Variation

Objective 1   Write an equation expressing direct variation.

Objective 2   Find the constant of variation, and solve direct variation problems.

Suppose y varies directly as x. Find the equation connecting y and x.

1. $y = 15$ when $x = 5$.

2. $y = 30$ when $x = 6$.

3. $y = 12$ when $x = 8$.

4. $y = 11$ when $x = 22$.

5. $y = 23$ when $x = 12$.

6. $y = 14$ when $x = 9$.

7. $y = 150$ when $x = 3$.

8. $y = 270$ when $x = 9$.

9. $y = 8$ when $x = 6$.

10. $y = 13.75$ when $x = 55$.

Solve each problem.

11. If r varies directly as t, and $r = 10$ when $t = 2$, find r when $t = 9$.

12. If q varies directly as p, and $q = 36$ when $p = 5$, find q when $p = 20$.

13. If x varies directly as y, and $x = 9$ when $y = 2$, find x when y is 7.

14. If z varies directly as x, and $z = 15$ when $x = 4$, find z when x is 11.

15. If m varies directly as p^2, and $m = 20$ when $p = 2$, find m when p is 5.

16. If a varies directly as b^2, and $a = 48$ when $b = 4$, find a when $b = 7$.

17. If y varies directly as the square of z, and $y = 8$ when $z = 6$, find y when $z = 9$.

18. If m varies directly as the square of p, and $m = 1$ when $p = 2$, find m when p is 7.

19. The circumference of a circle varies directly as the radius. A circle with a radius of 7 centimeters has a circumference of 43.96 centimeters. Find the circumference of the circle if the radius changes to 11 centimeters.

20. The pressure exerted by a certain liquid at a given point varies directly as the depth of the point beneath the surface of the liquid. The pressure at 10 feet is 50 pounds per square inch (psi). What is the pressure at 20 feet?

21. The force required to compress a spring varies directly as the change in length of the spring. If a force of 20 newtons is required to compress a spring 2 centimeters in length, how much force is required to compress a spring of length 10 centimeters?

22. The surface area of a sphere varies directly as the square of its radius. If the surface area of a sphere with a radius of 12 inches is 576π square inches, find the surface area of a sphere with a radius of 3 inches.

Objective 3 **Solve inverse variation problems.**

Suppose y varies inversely as x. Find the equation connecting y and x.

23. $y = 10$ when $x = 2$.

24. $y = 4$ when $x = 6$.

25. $y = 8$ when $x = 10$.

26. $y = 2$ when $x = 12$.

27. $y = 5$ when $x = \dfrac{1}{6}$.

28. $y = 9$ when $x = \dfrac{3}{2}$.

Solve each problem.

29. If y varies inversely as x, and $y = 10$ when $x = 3$, find y when $x = 12$.

30. If z varies inversely as y^2, and $z = 12$ when $y = 3$, find z when $y = 6$.

31. If p varies inversely as q^2, and $p = 4$ when $q = \frac{1}{2}$, find p when $q = \frac{3}{2}$.

32. If z varies inversely as x^2, and $z = 9$ when $x = \frac{2}{3}$, find z when $x = \frac{5}{4}$.

33. The current in a simple electrical circuit varies inversely as the resistance. If the current is 50 amperes (an ampere is a unit for measuring current) when the resistance is 10 ohms (an ohm is a unit for measuring resistance), find the current if the resistance is 5 ohms.

34. The illumination produced by a light source varies inversely as the square of the distance from the source. If the illumination produced 4 feet from a light source is 75 footcandles, find the illumination produced 9 feet from the same source.

35. The weight of an object varies inversely as the square of its distance from the center of the earth. If an object 8000 miles from the center of the earth weighs 90 pounds, find its weight when it is 12,000 miles from the center of the earth.

36. The speed of a pulley varies inversely as its diameter. One kind of pulley, with a diameter of 3 inches, turns at 150 revolutions per minute. Find the speed of a similar pulley with diameter of 5 inches.

Objective 4 **Solve joint variation problems.**

Suppose y varies jointly as x and z. Find the equation connecting y, x, and z.

37. $y = 10$ when $x = 2$ and $z = 5$.

38. $y = 50$ when $x = 4$ and $z = 6$.

39. $y = 27$ when $x = 9$ and $z = 6$.

40. $y = 144$ when $x = 8$ and $z = 9$.

41. $y = 24$ when $x = 12$ and $z = 4$.

42. $y = 30$ when $x = 10$ and $z = 12$.

Solve each problem.

43. If r varies jointly as m and n^2, and $r = 72$ when $m = 4$ and $n = 6$, find r when $m = 3$ and $n = 4$.

44. If q varies jointly as p and r^2, and $q = 27$ when $p = 9$ and $r = 2$, find q when $p = 8$ and $r = 4$.

45. Suppose y varies jointly as x^2 and z^2, and $y = 72$ when $x = 2$ and $z = 3$. Find y when $x = 4$ and $z = 2$.

46. Suppose d varies jointly as f^2 and g^2, and $d = 384$ when $f = 3$ and $g = 8$. Find d when $f = 6$ and $g = 2$.

47. For a fixed period of time, interest varies jointly as the principal and the rate of interest. If a principal of $2000 invested at a rate of 6% earns $300 in interest, how much interest will a principal of $4000 earn if it is invested at 8% for the same period of time?

48. The absolute temperature of an ideal gas varies jointly as its pressure and its volume. If the absolute temperature is $250°$ when the pressure is 25 pounds per square centimeter and the volume is 50 cubic centimeters, find the absolute temperature when the pressure is 50 pounds per square centimeter and the volume is 75 cubic centimeters.

Objective 5   **Solve combined variation problems.**

Suppose y varies directly as x and inversely as z. Find the equation connecting y, x, and z.

49. $y = 1$ when $x = 2$ and $z = 6$.

52. $y = 3$ when $x = 72$ and $z = 8$.

50. $y = 6$ when $x = 4$ and $z = 8$.

53. $y = .1625$ when $x = 4$ and $z = 16$.

51. $y = 2$ when $x = 6$ and $z = 2$.

54. $y = 10.5$ when $x = 1.8$ and $z = .6$.

Solve each problem.

55. The time required to lay a sidewalk varies directly as its length and inversely as the number of people who are working on the job. If three people can lay a sidewalk 100 feet long in 15 hours, how long would it take two people to lay a sidewalk 40 feet long?

56. When an object is moving in a circular path, the centripetal force varies directly as the square of the velocity and inversely as the radius of the circle. A stone that is whirled at the end of a string 50 centimeters long at 900 centimeters per second has a centripetal force of 3,240,000 dynes. Find the centripetal force if the stone is whirled at the end of a string 75 centimeters long at 1500 centimeters per second.

7.6   Mixed Exercises

Solve each problem.

57. If p varies directly as q, and $p = 14$ when $q = 2$, write an equation connecting p and q.

58. If w varies jointly as v and s, and $w = 3.5$ when $v = 25$ and $s = 7$, write an equation connecting w, v, and s.

59. If r varies inversely as s, and $r = 7$ when $s = 8$, find r when $s = 12$.

60. If m varies directly as w and inversely as r^2, and $m = 1845$ when $w = 4.5$ and $r = .1$, find m when $w = 2.5$ and $r = .2$.

61. If r varies inversely as t^2, and $r = 8$ when $t = 4$, find r when $t = 9$.

62. p varies jointly as P, V, and t and inversely as v and T. Suppose $p = 65.625$ when $P = 50$, $V = 9$, $t = 350$, $v = 8$, and $T = 300$. Find p when $P = 60$, $V = 8$, $t = 300$, $v = 6$, and $T = 200$.

63. For a body falling free from rest (disregarding air resistance), the distance the body falls varies directly as the square of the time. If an object is dropped from the top of a tower 400 feet high and hits the ground in 5 seconds, how far did it fall in the first 3 seconds?

64. The resistance in ohms of a platinum wire temperature sensor varies directly as the temperature in degrees Kelvin (K). If the resistance is 646 ohms at a temperature of 190 K, find the resistance at a temperature of 250 K.

65. The volume of a gas varies inversely as the pressure and directly as the temperature. If a certain gas occupies a volume of 1.3 liters at 300 K and a pressure of 18 kilograms per square centimeter, find the volume at 340 K and a pressure of 24 kilograms per square centimeter.

66. The force required to compress a spring varies directly as the change in the length of the spring. If a force of 12 pounds is required to compress a certain spring 3 inches, how much force is required to compress the spring 5 inches?

67. The distance that a person can see to the horizon on a clear day from a point above the surface of the earth varies directly as the square root of the height at the point. If a person 144 meters above the surface can see 18 kilometers to the horizon, how far can a person see to the horizon from a point 64 meters above the surface?

68. For a fixed interest rate, interest varies jointly as the principal and the time in years. If a principal of $5000 invested for 4 years earns $900, how much interest will $6000 invested for 3 years earn at the same interest rate?

Chapter 8

EQUATIONS, INEQUALITIES, AND SYSTEMS REVISITED

8.1   Review of Solving Linear Equations and Inequalities

Objective 1   Solving linear equations.

Decide whether or not the number is a solution of the given equation.

1. $5k = 35$; 7

2. $r + 12 = 10$; 2

3. $6m = 9$; $\dfrac{3}{2}$

4. $-s + 6 = 8$; -2

5. $\dfrac{3}{4}y = -12$; -16

6. $13 - k = 11$; -2

7. $6 - \dfrac{1}{2}p = 8$; -4

8. $\dfrac{1}{3}q - \dfrac{5}{3} = \dfrac{7}{3}$; 4

9. $8 = x - 4$; 12

10. $-22 = 6k - 4$; -3

Solve and check each equation.

11. $3x + 6 + 2x = 4x - 3$

12. $7r - 4r + 2 = r - 4$

13. $12a + 5 = 5a - 2$

14. $6z - 3z + 5 - 4 = 3z - 5 + 4$

15. $11x - 14x - 7 + 8 = 4x + 5 - 2$

16. $9r - 4r + 8r - 6 = 10r - 11 + 2r$

17. $9x - 4 + 6x - 1 = 12 - 2x$

18. $18 - 3y + 11 = 7y + 6y - 27 - 9y$

19. $7t - 11t - 6 + 15 = -t + 17$

20. $21w - 14 + 8 - 26w = 3w - 18$

Solve and check each equation.

21. $2(x + 4) = 3x - 2$

22. $3(x - 2) = 7x + 2$

23. $2(z - 1) + 3(z - 4) = 5$

24. $5(2p + 1) - (p + 3) = 7$

25. $-11z - (5 - 6z) = -(6 - 3z) + 1$

26. $-9w - (4 + 3w) = -(2w - 1) - 5$

27. $4t - 3(4 - 2t) = 2(t - 3) + 6t + 2$

28. $-[p - (4p + 2)] = 3 + (4p + 7)$

29. $6 - 5(2 - 3r) + 7 = 4(r - 6) - 17$

30. $-11a - (6a - 4) = -(5 - 3a) + 1$

31. $k + 2(-k + 4) - 3(k + 5) = -4$

32. $-r + 9(r - 5) + 4(3r + 4) = r - 5$

33. $4m - 3(5 - 2m) = 6(m - 3) + 2m + 1$

34. $6x - (8 - x) = 9[-2 - (5 + 2x) - 12]$

Solve and check each equation.

35. $\dfrac{1}{3}m = 17$

43. $\dfrac{p-2}{3} + \dfrac{p}{4} = \dfrac{1}{2}$

36. $-\dfrac{5}{7}y = 14$

44. $\dfrac{y-8}{5} + \dfrac{y}{3} = -\dfrac{12}{5}$

37. $-\dfrac{4}{5}x = 8$

45. $\dfrac{2x+5}{5} - \dfrac{3x+1}{2} = \dfrac{7-x}{2}$

38. $\dfrac{z}{2} + \dfrac{z}{3} = 5$

46. $\dfrac{x-5}{2} - \dfrac{x+6}{3} = -4$

39. $-\dfrac{4}{3}x + 5 = 7$

47. $.06x + .14(x + 500) = 130$

48. $.35(140) + .15w = .05(w + 1100)$

40. $-\dfrac{4}{7}q - 2 = 6$

49. $.07(w - 14) - .05w = 3.22$

50. $.08(x - 9) = 1.38 - .02x$

41. $\dfrac{6w}{7} = 24$

51. $5.24 - .06(x + 9) = .04x$

52. $.04(x + 6) - .02x = 3.16$

42. $\dfrac{m}{6} + \dfrac{m}{4} = -9$

Objective 2   Identify conditional equations, contradictions, and identities.

Decide whether each equation is a conditional equation, *an* identity, *or a* contradiction. *Give the solution set.*

53. $5m - 2(4 - m) = 6$

58. $2(2y - 5) - 3(4 - y) = 7y - 22$

54. $2[3 - 4(5 - r)] = 2(3r - 11)$

59. $28 - 4k = 36 + 2(2k - 4)$

55. $6t - 3(5 + 2t) = -12$

60. $7(3 - 4q) - 10(q - 2) = 19(5 - 2q)$

56. $7(2 - 5b) - 32 = 10b - 3(6 + 15b)$

61. $6z - 3(8z - 4) = 2(4 - 7z) - 4(z - 1)$

57. $6(3 - 4x) + 10 = -15x + 3(2 - 3x)$

62. $13p - 9(3 - 2p) = 3(10p - 9) + 1$

Objective 3   Solve linear inequalities.

Solve each inequality, giving its solution set in interval form.

63. $x + 7 \le 4$

66. $t - 2 \ge 7$

69. $8s < 7s - 3$

64. $p - 3 > 6$

67. $z + 8 < 8$

70. $x - 3 \ge -5$

65. $x + 5 > -1$

68. $6s < 7s - 3$

71. $a + 5 > 6$

72. $y - 7 \leq 8$

73. $p + 2 > 6$

74. $6 < -3t + 4t$

Solve each inequality, giving its solution set in both interval and graph forms. Check your answers.

75. $2r \leq 6$

76. $-3a < 9$

77. $2z < -8$

78. $-2x \geq 5$

79. $-16z \leq -64$

80. $-\dfrac{1}{2}k \geq 5$

81. $-\dfrac{3}{4}r \geq 27$

82. $-\dfrac{3}{5}x \leq -6$

83. $-\dfrac{2}{7}x > -4$

84. $20 - 3(2p + 4) \leq -10p$

85. $14 - 3(p + 2) < -5p$

86. $8 - 4(p - 2) > -6p$

87. $12 - 2(p - 3) \geq -8p$

Objective 4   **Solve three-part inequalities.**

Solve each inequality, giving its solution set in both interval and graph forms. Check your answers.

88. $-2 < y - 3 < 6$

89. $-6 < k + 2 < 8$

90. $10 < z + 5 < 14$

91. $-4 \leq a + 5 < -2$

92. $-4 < 6 - 2x < -2$

93. $4 > 3a + 4 > -4$

94. $-3 \leq \dfrac{2t + 1}{6} \leq 5$

95. $-3 \leq \dfrac{6q - 1}{4} \leq 0$

96. $-5 \leq \dfrac{3}{4}x + 1 \leq 10$

97. $-4 \leq \dfrac{2}{5}x + 2 \leq 6$

8.1   Mixed Exercises

Decide whether or not the given number is a solution of the equation.

98. $3y - 8y = 45; \ 5$

99. $15 - 7x = 12; \ \dfrac{3}{7}$

Solve and check each equation.

100. $5x - 7x + 14 = 4 - 11 - 9x$

101. $8(p - 4) = 6p - 10$

102. $9q + 114 - 2q = 15 - 6q + 8$

103. $6a - 3(5a + 2) = 4 - 5a$

104. $3(x - 6) - 4(8 - 3x) = 10(x - 4) + 4$

105. $5(1 - 2x) + 3(x - 4) = -7(1 + x)$

106. $\dfrac{2}{3}k = 8$

107. $-\dfrac{5}{7}y = 15$

108. $\dfrac{a + 8}{6} + \dfrac{4a}{9} = \dfrac{2a + 12}{9}$

109. $\dfrac{2a + 3}{5} - \dfrac{3a - 1}{2} = \dfrac{4a + 7}{2}$

110. $4(x - 12) - 8(x + 1) = 2(28 - 2x)$

111. $3(t - 4) + 5(6 - 2t) = 7(2 - t) + 4$

112. $.05(x - 2) + .06x = .78$

113. $.08x - .05(x - 94) - 4.88 = 0$

Solve each inequality, giving its solution set in both interval and graph forms.

114. $-\dfrac{2}{5}x < \dfrac{3}{5}$

115. $-4 \le \dfrac{1 - 2x}{6} \le 0$

116. $4 < 3m + 5 < 7$

117. $-3 < 2t + 4 < 6$

118. $-3 \le \dfrac{3y - 1}{-4} \le 1$

119. $-\dfrac{4}{7}y \ge \dfrac{3}{14}$

8.2   Set Operations and Compound Inequalities

Objective 1   Find the intersection of two sets.

Find each intersection of sets.

1. $\{0, 1, 2, 3\} \cap \{2, 3, 4, 5\}$

2. $\{-6, -5, -4\} \cap \{-3, -2, -1\}$

3. $\{7, 8, 9, 10\} \cap \varnothing$

4. $\{2, 6, 8\} \cap \{2, 6, 8\}$

Let A = {0, 1, 2, 3, 4, 5}, B = {2, 4, 6, 8, 10}, C = {1, 3, 5, 7, 9}, D = {0, 2, 4}, and E = {0}. Specify each intersection.

5. $A \cap B$

6. $A \cap C$

7. $A \cap D$

8. $A \cap E$

9. $B \cap C$

10. $B \cap D$

Objective 2   Solve compound inequalities with the word and.

For each compound inequality, give the solution set in both interval and graph forms.

11. $r < 3$ and $r > 0$

12. $m \le 4$ and $m \le 7$

13. $t \ge -2$ and $t \ge 1$

14. $x - 3 \le 6$ and $x + 2 \ge 7$

15. $2q < -2$ and $q + 3 > 1$

16. $2z + 1 < 3$ and $3z - 3 > 3$

17. $3x + 2 < 11$ and $2 - 3x \le 14$

18. $5t > 0$ and $5t + 4 \le 9$

19. $q < -1$ and $q \ge 2$

20. $r \ge 2$ and $r \le -2$

Objective 3   Find the union of two sets.

Find each union of sets.

21. $\{0, 1, 2, 3\} \cup \{2, 3, 4, 5\}$

22. $\{-6, -5, -4\} \cup \{-3, -2, -1\}$

23. $\{7, 8, 9, 10\} \cup \varnothing$

24. $\{2, 6, 8\} \cup \{2, 6, 8\}$

Let A = {1, 2, 3, 4, 5, 6}, B = {0, 2, 4, 6, 8, 10}, C = {1, 3, 5, 7, 9}, D = {1, 2, 3}, and E = {0}. Specify each union.

25. $A \cup B$

26. $A \cup C$

27. $A \cup D$

28. $A \cup E$

29. $B \cup C$

30. $B \cup D$

Objective 4   Solve compound inequalities with the word or.

For each compound inequality, give the solution set in both interval and graph forms.

31. $m > 4$ or $m < -1$

32. $y \le 1$ or $y \ge 6$

33. $a \le 2$ or $a \ge 6$

34. $k \le 2$ or $k \ge 6$

35. $r \ge -1$ or $r \ge 4$

36. $p \ge -1$ or $p \le 6$

37. $q + 3 > 7$ or $q + 1 \leq -3$

38. $s - 5 > 0$ or $s + 7 < 6$

39. $4x < x - 5$ or $6x > 2x + 3$

40. $3 > 4m + 2$ or $7m - 3 \geq -2$

8.2 Mixed Exercises

Let A = {0}, B = {2, 4}, C = {1, 3, 5, 9}, and D = {0, 1, 2, 3, 4, 5}. **Specify each set.**

41. $A \cup B$ **42.** $C \cap D$ **43.** $A \cup D$ **44.** $B \cap C$

For each compound inequality, give the solution set in both interval and graph forms.

45. $z \geq 0$ or $z \leq -2$

46. $y \geq -2$ and $y < 3$

47. $2r + 4 \geq 8$ or $4r - 3 < 1$

48. $4t < 2t + 10$ or $t - 3 > 3$

49. $z \leq 2$ and $z \geq 6$

50. $x \geq -4$ or $x \leq 5$

51. $-2x + 1 < 3$ and $3x \leq 12$

52. $7y + 5 \leq 3$ and $-3y \geq -9$

8.3   Absolute Value Equations and Inequalities

Objective 1 Use the distance definition of absolute value.

Graph the solution set of each equation or inequality.

1. $|m| = 7$

2. $|q| = 0$

3. $|k| < 8$

4. $|r| > 2$

5. $|x| \geq 6$

6. $|y| \geq 0$

7. $|p| \geq -2$

8. $|p| < 3$

9. $|x| \leq 10$

10. $|t| \leq 0$

Objective 2 Solve equations of the form $|ax + b| = k$, for $k > 0$.

Solve each equation.

11. $|t - 4| = 7$

12. $|m + 4| = 8$

13. $|3 - q| = 7$

14. $|3k - 1| = 6$

15. $|5 - t| = 3$

16. $|m + 6| = 2$

17. $|2x + 3| = 10$

18. $|2r + 3| = 0$

19. $|5r - 15| = 0$

20. $\left| 5 - \dfrac{4}{3}x \right| = 9$

Objective 3 Solve inequalities of the form $|ax + b| < k$ and of the form $|ax + b| > k$, for $k > 0$.

Solve each inequality and graph the solution set.

21. $|x - 2| > 8$

22. $|q + 5| > 15$

23. $|n + 5| < 8$

24. $|5 - z| < 8$

25. $|5r + 2| < 18$

26. $|2r - 9| \geq 23$

27. $|3q - 5| + 2 \geq 6$

28. $|-k - 3| \geq 1$

29. $|2z + 4| + 6 > 8$

30. $|p - 5| - 5 \geq 0$

31. $|2 - z| \leq 3$

32. $|4y - 1| - 3 \leq -1$

Objective 4 Solve absolute value equations that involve rewriting.

Solve each equation.

33. $|a|+5=7$

34. $|z|-6=3$

35. $|s|+2=0$

36. $|y|-5=-7$

37. $|5+y|+3=7$

38. $|7t+5|+6=14$

39. $|5-2w|+7=5$

40. $|2w-1|+7=12$

41. $\left|2-\dfrac{1}{2}x\right|-5=18$

42. $|4t+3|+8=10$

$\boxed{\textbf{Objective 5}}$ **Solve equations of the form** $|ax+b|=|cx+d|$.

Solve each equation.

43. $|2x+6|=|3x-9|$

44. $|a-4|=|a-3|$

45. $|y+3|=|2y-5|$

46. $|5-z|=|2z+3|$

47. $|3-a|=|a+5|$

48. $|y+5|=|3y+1|$

49. $|2x-8|=|6x+7|$

50. $|2p-4|=|7-p|$

51. $\left|p-\dfrac{1}{2}\right|=\left|\dfrac{1}{2}p-1\right|$

52. $\left|y-\dfrac{1}{4}\right|=\left|\dfrac{1}{2}y+1\right|$

$\boxed{\textbf{Objective 6}}$ **Solve special cases of absolute value equations and inequalities.**

Solve each equation.

53. $|2x-4|=-6$

54. $\left|7+\dfrac{1}{2}x\right|=0$

55. $|p|=0$

56. $|a|\le-2$

57. $|k+5|\le-2$

58. $|3-2x|+5\le1$

59. $|4+t|<0$

60. $|2+q|<0$

61. $|m-2|\ge-1$

62. $|3p+4|>-7$

8.3   Mixed Exercises

Solve each equation.

63. $|2y+5|=3$

64. $\left|\dfrac{1}{2}x-3\right|=4$

65. $|x+7|=-3$

66. $|r+5|=6$

67. $|r|+7=5$

68. $|3t+2|-9=18$

69. $|3x-2|=|5x+8|$

70. $\left|\dfrac{2}{3}z+1\right|=\left|\dfrac{1}{3}z-1\right|$

Solve each inequality and graph the solution set.

71. $|3-2a|<7$

72. $|3r-9|\le10$

73. $|2x+7|>9$

74. $|3-y|\ge7$

75. $|2z+1|+4\ge15$

76. $|4x|+3>5$

77. $|3y-1|+3\le5$

78. $|4n-3|+6<11$

8.4   Review of Systems of Linear Equations in Two Variables

Objective 1   Solve linear systems (with two equations and two variables).

Solve each system by graphing.

1. $x + y = 3$
   $x - y = -1$

2. $x + y = 5$
   $x - y = 3$

3. $x - y = -2$
   $3x + 2y = -6$

4. $x + 4y = 1$
   $2x + y = 2$

5. $x + 2y = 4$
   $3x + y = -3$

6. $x - 2y = 1$
   $x + 4y = 7$

7. $x + 2y = 0$
   $2x + y = -6$

8. $x + y = 2$
   $2x + 5y = 10$

Decide whether the ordered pair is a solution for the given system.
Write **solution** *or* **not a solution.**

9. $2x - 5y = 5$ $(5, 1)$
   $4x + 3y = 23$

10. $4x + 5y = 11$ $(-1, 3)$
    $3x - y = -6$

11. $4x - 5y = 26$ $(4, -2)$
    $3x + 2y = 6$

12. $5x - 3y = -19$ $(-2, 3)$
    $2x + 3y = 6$

13. $7x + 9y = 15$ $(6, -3)$
    $4x + 8y = 21$

14. $3x - y = 7$ $(3, 2)$
    $4x - 5y = 2$

15. $x = 4y$ $(12, 3)$
    $3x = y + 33$

16. $5y = x$ $(15, 3)$
    $2y = x - 9$

17. $2x - 9y = 27$ $(0, -3)$
    $5x + 6y = 18$

18. $7x + 8y = -14$ $(-2, 0)$
    $5x - 7y = 10$

Solve each system by the substitution method.

19. $3x + y = -20$
    $y = 2x$

20. $4x - y = -7$
    $y = -3x$

21. $x + 2y = 5$
    $x = 2y + 1$

22. $4x - 3y = 15$
    $x = y + 4$

23. $3x + 7y = 16$
    $y = 2x - 5$

24. $2x + y = 6$
    $y = 5 - 3x$

25. $y = 11 - 2x$
$x = 18 - 3y$

26. $y = 2x + 5$
$y = -4x + 2$

27. $x + 6y = 15$
$y = \dfrac{2}{3}x$

28. $3x - 2y = -1$
$x = \dfrac{3}{4}y$

29. $\dfrac{x}{2} + \dfrac{y}{3} = \dfrac{3}{2}$
$y = 3x$

30. $\dfrac{x}{4} - \dfrac{y}{5} = \dfrac{3}{4}$
$y = 5x$

31. $3x + 2y = 7$
$4x - 3y = -19$

32. $4x + 5y = 13$
$3x - 4y = 2$

33. $x + 3y = 9$
$x - 2y = -1$

Use the elimination method to solve each system of linear equations.

34. $x + 2y = 7$
$x - y = -2$

35. $3x - y = 11$
$x + y = 5$

36. $5x + 2y = 16$
$3x - 4y = 20$

37. $3x + 4y = -13$
$5x - 2y = -13$

38. $8x + 3y = 13$
$3x + 2y = 11$

39. $2x + 8y = 3$
$4x - 12y = -1$

40. $3x + 4y = 3$
$9x - 8y = 4$

41. $\dfrac{1}{2}x + \dfrac{1}{4}y = 5$
$\dfrac{1}{2}x - \dfrac{3}{4}y = -3$

42. $\dfrac{x}{5} + \dfrac{y}{2} = 0$
$\dfrac{3x}{5} - \dfrac{5y}{2} = -8$

43. $\dfrac{x}{2} + \dfrac{y}{3} = -2$
$\dfrac{3x}{2} + \dfrac{5y}{3} = -8$

Objective 2   **Solve special systems.**

44. $2x + y = 3$
$4x + 2y = 12$

45. $3x - y = 8$
$-6x + 2y = 16$

46. $5x - 2y = 4$
$10x - 4y = 8$

47. $9x - y = 6$
$-18x + 2y = -12$

48. $4x - 3y = 6$
$8x - 6y = 10$

8.4   Mixed Exercises

Solve each system by any method.

49. $8x + 7y = 52$
$x - 3 = 0$

59. $5x + 15y = 5$
$x + 3y = 18$

50. $6x + 5y = 13$
$y + 1 = 0$

60. $-x + 2y = -8$
$2x - 4y = 16$

51. $3x + 2y = -8$
$4 + y = 2x$

52. $3x - 4y = 13$
$2x + 3y = 3$

53. $\dfrac{x}{6} + \dfrac{y}{4} = 1$
$4x + 6y = 24$

54. $3x + 7y = 32$
$y = \dfrac{1}{3}x$

55. $2x + 3y = 1$
$3x - 2y = 8$

56. $4x - 3y = -12$
$x + 3 = y$

57. $6x - 5y = -8$
$x = \dfrac{1}{2}y$

58. $\dfrac{x}{5} - \dfrac{y}{3} = \dfrac{3}{5}$
$-3x + 5y = 9$

8.5   Systems of Linear Equations in Three Variables; Applications

Objective 2   Solve linear systems (with three equations and three variables) by elimination.

Solve each system of equations.

1. $x + y + z = 0$
   $x - y + z = -2$
   $x - y - z = -4$

2. $x + 2y + z = 4$
   $2x + y - z = -1$
   $-x + y + z = 2$

3. $x + y + z = 4$
   $x - y + 2z = 8$
   $2x + y - z = 3$

4. $3x - y + 2z = -6$
   $2x + y + 2z = -1$
   $3x + y - z = -10$

5. $4x + 2y + 3z = 11$
   $2x + y - 4z = -22$
   $3x + 3y + z = -1$

6. $x - 2y + 5z = -7$
   $2x + 3y - 4z = 14$
   $3x - 5y + z = 7$

7. $2x - 5y + 2z = 30$
   $x + 4y + 5z = -7$
   $\dfrac{1}{2}x - \dfrac{1}{4}y + z = 4$

8. $5x - 2y + z = 28$
   $3x + 5y - 2z = -23$
   $\dfrac{2}{3}x + \dfrac{1}{3}y + z = 1$

9. $\dfrac{1}{3}x + \dfrac{1}{6}y - \dfrac{2}{3}z = -1$
   $\dfrac{3}{4}x + \dfrac{1}{3}y + \dfrac{1}{4}z = -3$
   $\dfrac{1}{2}x + \dfrac{3}{2}y + \dfrac{3}{4}z = 21$

10. $\dfrac{2}{3}x - \dfrac{1}{4}y + \dfrac{5}{8}z = 0$
    $\dfrac{1}{5}x + \dfrac{2}{3}y - \dfrac{1}{4}z = -7$
    $\dfrac{3}{5}x - \dfrac{4}{3}y + \dfrac{7}{8}z = 5$

Objective 3   Solve linear systems where some of the equations have missing terms.

Solve each system of equations.

11. $x - z = -3$
    $y + z = 4$
    $x - y = 3$

12. $2x + 3y = 3$
    $6y - 5z = 3$
    $4x + 9y = 8$

13. $x + 5y = -23$
    $4y - 3z = -29$
    $2x + 5z = 19$

14. $3x - 4z = -23$
    $y + 5z = 24$
    $x - 3y = 2$

15. $4x - 5y = -13$
    $3x + z = 9$
    $2y + 5z = 10$

16. $7x + z = 5$
    $3y - 2z = -16$
    $5x + y = -2$

17. $2x + 5y \quad = 18$
$\qquad 3y + 2z = 4$
$\dfrac{1}{4}x - y \quad = -1$

18. $5x - \quad 2z = 8$
$\qquad 4y + 3z = -9$
$\dfrac{1}{2}x + \dfrac{2}{3}y \quad = -1$

19. $x + 2y \quad = -2$
$\qquad \dfrac{1}{2}y + z = -1$
$\dfrac{2}{3}x - \dfrac{3}{4}y \quad = 7$

20. $4x - \quad z = -6$
$\qquad \dfrac{3}{5}y + \dfrac{1}{2}z = 0$
$\dfrac{1}{3}x + \quad \dfrac{2}{3}z = -5$

Objective 4   **Solve special systems.**

Solve each systems of equations.

21. $x - y + z = 7$
$2x + 5y - 4z = 2$
$-x + y - z = 4$

22. $8x - 7y + 2z = 1$
$3x + 4y - z = 6$
$-8x + 7y - 2z = 5$

23. $3x - 2y + 4z = 5$
$-3x + 2y - 4z = -5$
$\dfrac{3}{2}x - y + 2z = \dfrac{5}{2}$

24. $-x + 5y - 2z = 3$
$2x - 10y + 4z = -6$
$-3x + 15y - 6z = 9$

25. $8x - 4y + 2z = 0$
$3x + y - 4z = 0$
$5x + y + 2z = 0$

26. $x - 3y + 4z = 0$
$2x + y - z = 0$
$-x + y - 5z = 0$

27. $3x - 2y + 5z = 6$
$x - 4y - z = 1$
$\dfrac{3}{2}x - y + \dfrac{5}{2}z = -3$

28. $2x + 7y - 8z = 3$
$5x - y - z = 1$
$x + \dfrac{7}{2}y - 4z = 3$

29. $x - 5y + 2z = 0$
$-x + 5y - 2z = 0$
$\dfrac{1}{2}x - \dfrac{5}{2}y + z = 0$

30. $3x - 2y + 5z = 0$
$6x - 4y + 10z = 0$
$\dfrac{3}{2}x - y + \dfrac{5}{2}z = 0$

| Objective 5 | Solve application problems with three variables using a system of three equations. |

For each problem, select variables to represent the three unknowns, write three equations using the three variables, and solve the resulting system.

31. Three numbers have a sum of 31. The middle number is 1 more than the smallest number. The sum of the smaller two numbers is 7 more than the largest number. Find the three numbers.

32. The sum of three numbers is 99. The difference of the smaller two is 3. The sum of the smallest and twice the largest is 108. Find the three numbers.

33. The sum of the measures of the angles of any triangle is 180°. In a certain triangle, the first angle measures 20° less than the second angle, and the second angle measures 10° more than the third. Find the measures of the three angles.

34. In a certain triangle, the sum of the measures of the smallest and largest angles is 50° more than the measure of the medium angle. The medium angle measures 25° more than the smallest angle. Find the measures of the three angles.

35. The perimeter of a triangle is 60 inches. The second side is 10 inches longer than the first side, and the third side is 6 inches more than twice the first side. Find the lengths of the sides.

36. Lance has some $5, $10, and $20-bills. He has a total of 51 bills, worth $795. The number of $5-bills is 25 less than the number of $20-bills. Find the number of each type of bill he has.

37. Emily has some $10, $20, and $50-bills. She has a total of 50 bills, worth $1270. The number of $20 bills equals the combined number of $10 and $50-bills. Find the number of each type of bill.

38. Sara Mitchell has $80,000 to invest. She invests part at 5%, one fourth this amount at 6%, and the balance at 7%. Her total annual income from interest is $4700. Find the amount invested at each rate.

39. A company borrowed a total of $75,000. Some of the money was borrowed at 8% interest, and $30,000 more than that amount was borrowed at 10%. The rest was borrowed at 11%. How much was borrowed at each rate if the total annual simple interest was $7150?

40. A merchant wishes to mix gourmet coffee selling for $8 per pound, $10 per pound, and $15 per pound to get 50 pounds of a mixture that can be sold for $11.70 per pound. The amount of the $8 coffee must be 3 pounds more than the amount of the $10 coffee. Find the number of pounds of each that must be used.

8.5   Mixed Exercises

Solve each system of equations.

41. $x + y + z = 6$
    $2x - y + z = 3$
    $x + 2y - z = 2$

42. $3x - y + 2z = 9$
    $2x + y - z = 7$
    $x + 2y - 3z = 4$

43. $2x + y - 4z = 17$
    $3x - 2y - z = -7$
    $-2x + 2y + z = 7$

44. $6x + 2y - 4z = 6$
    $-12x - 4y + 8z = -12$
    $-3x - y + 2z = -3$

45. $\begin{aligned} 2x - 3y &= -2 \\ x + y - 4z &= -16 \\ 3x - 2y + z &= 7 \end{aligned}$

46. $\begin{aligned} 9x - 6z &= 8 \\ 3x + 4y - 2z &= 16 \\ -6x + 4z &= 11 \end{aligned}$

47. $\begin{aligned} 2x + 5y + 3z &= -1 \\ 3x - y + 2z &= 7 \\ 4x + 2y + 3z &= 6 \end{aligned}$

48. $\begin{aligned} 5x - 2y + 5z &= 27 \\ 4x + 3y + 4z &= 17 \\ 2x - 4y - 3z &= -1 \end{aligned}$

49. $\begin{aligned} 3x - y &= -2 \\ y + 5z &= -4 \\ -2x + 3y - z &= -8 \end{aligned}$

50. $\begin{aligned} 3x - 4y + 2z &= 3 \\ x + 3y + 5z &= 14 \\ 2x - 5y + 4z &= 5 \end{aligned}$

Use a system of equations to solve the problem.

51. The side of a square is 5 centimeters shorter than the side of an equilateral triangle. The perimeter of the square is 7 centimeters less than the perimeter of the triangle. Find the lengths of a side of the square and of a side of the triangle.

8.6   Solving Systems of Linear Equations by Matrix Methods

Objective 1   **Define a matrix.**

Give the dimensions of each matrix.

1. $\begin{bmatrix} 1 & 2 \\ -1 & 0 \end{bmatrix}$

2. $\begin{bmatrix} -4 & 3 \\ 0 & 7 \\ 6 & 2 \end{bmatrix}$

3. $\begin{bmatrix} 1 & 4 & 7 \\ 6 & 5 & -5 \end{bmatrix}$

4. $\begin{bmatrix} 1 & 4 \\ 3 & 2 \\ -2 & 0 \\ 5 & -3 \end{bmatrix}$

5. $\begin{bmatrix} 3 & 2 & 6 \\ 6 & 0 & 1 \\ -2 & 10 & -11 \\ 1 & 5 & 2 \end{bmatrix}$

6. $\begin{bmatrix} 8 & 7 & 6 & 4 \\ 4 & -1 & 0 & 6 \\ -5 & 3 & -4 & 7 \end{bmatrix}$

Objective 2   **Write the augmented matrix for a system.**

Write the augmented matrix for each system.

7. $3x - 4y = 7$
   $2x + y = 12$

8. $2x - 3y = 12$
   $7x + 3y = 15$

9. $\dfrac{1}{3}x - \dfrac{1}{2}y = 7$
   $\dfrac{5}{3}x + \dfrac{1}{2}y = 8$

10. $\dfrac{1}{2}x + \dfrac{1}{2}y = -16$
    $-3x + y = 2$

11. $3x = y + 4$
    $y = 5x - 2$

12. $y + 3 = -2x$
    $x = 4y - 5$

13. $-2x + 3y - 5z = 7$
    $6x + 2y - 4z = 12$
    $5x - 2y + z = -1$

14. $x + y + z = 10$
    $2x + y - 3z = 11$
    $x + 2z = -2$

Objective 3   Use row operations to solve a system with two equations.

Use row operations to solve each system.

15. $x + 3y = -7$
     $4x + 3y = -1$

16. $x - 2y = -1$
     $2x + y = 8$

17. $3x - 3y = 15$
     $2x + y = 4$

18. $x - 3y = 6$
     $2x + 3y = -6$

19. $3x - 2y = -6$
     $2x - y = -3$

20. $2x - y = 3$
     $3x + y = 2$

21. $y = x - 2$
     $2x = -3y + 9$

22. $x = 3y + 4$
     $2y = 5x + 19$

23. $3x + 2y = 12$
     $5x - 2y = 12$

Objective 4   Use row operations to solve a system with three equations.

Use row operations to solve each system.

24. $x - y - z = 6$
     $-x + 3y + 2z = -11$
     $3x + 2y + z = 1$

25. $x + y + z = 5$
     $x - 2y + 3z = 16$
     $2x - y + z = 9$

26. $x + 2y + 3z = 1$
     $x + y + 2z = 0$
     $2x - y - z = 1$

27. $x + y + z = 6$
     $x + 2y - 3z = -11$
     $-2x + y - z = -11$

28. $2x + y + z = 8$
     $x - y + z = 3$
     $3x + y - z = 1$

29. $2x - y + 2z = -1$
     $-x - 3y + z = 1$
     $x + y + z = 1$

Objective 5   Use row operations to solve special systems.

Use row operations to solve each system.

30. $x + y = 3$
     $3x + 3y = -2$

31. $x - 3y = 1$
     $4x - 12y = 5$

32. $x - 2y = 3$
     $3x - 6y = 9$

33. $2x + y = 10$
     $-4x - 2y = -20$

34. $x + y + z = 6$
     $x - y - z = 2$
     $2x + y + z = 8$

35. $x - 2y + 3z = 1$
$2x - y + 3z = 2$
$4x - y + 5z = 5$

37. $x + 2y - z = 6$
$2x + 4y - 2z = 12$
$-3x - 6y + 3z = -18$

36. $x + 3y - z = 1$
$2x + y - z = 2$
$4x - 3y - z = 4$

8.6 Mixed Exercises

Give the dimensions of each matrix.

38. $\begin{bmatrix} 2 & 1 & -1 \\ 0 & 4 & 5 \end{bmatrix}$

39. $\begin{bmatrix} 4 & 8 & -10 \\ -8 & 0 & 4 \\ 5 & -1 & 2 \\ 7 & 3 & -4 \end{bmatrix}$

Write the augmented matrix for each system.

40. $2x - 5y = 2$
$3x + 4y = -1$

41. $y = -2x + 7$
$x = 4y$

Use row operations to solve each system.

42. $x - 3y = -9$
$\dfrac{1}{3}x - y = 2$

43. $x - 2y = 4$
$-2x + y = 1$

44. $x - 3y = 10$
$2x + y = 6$

45. $2x - y = 2$
$6x - 3y = 6$

46. $x - 2y + z = 4$
$3x + 3y - 3z = -1$
$4x + y - 2z = 1$

47. $x - 4y - z = 6$
$2x - y + 3z = 0$
$3x - 2y + z = 4$

48. $2x - 2y - z = -16$
$-2x + 2y + 3z = 20$
$x + y + z = 1$

Chapter 9

ROOTS, RADICALS, AND ROOT FUNCTIONS

9.1   Radical Expressions and Graphs

Objective 1   Find square roots.

Find each root that is a real number. Use a calculator as needed.

1. $\sqrt[3]{27}$

2. $\sqrt[3]{216}$

3. $\sqrt[3]{-8}$

4. $\sqrt[4]{16}$

5. $\sqrt[4]{-81}$

6. $\sqrt[5]{243}$

7. $\sqrt[3]{-125}$

8. $\sqrt[5]{32}$

9. $\sqrt[4]{625}$

10. $\sqrt[5]{-1}$

11. $\sqrt[3]{3375}$

12. $\sqrt[3]{-64}$

13. $\sqrt[6]{64}$

14. $\sqrt[6]{729}$

15. $\sqrt[4]{256}$

16. $\sqrt[3]{-343}$

17. $\sqrt[4]{6561}$

18. $\sqrt[3]{512}$

19. $\sqrt[3]{-1331}$

20. $\sqrt[4]{14,641}$

Objective 2   Decide whether a given root is rational, irrational, or not a real number.

Objective 3   Find higher roots.

Find each root that is a real number.

21. $\sqrt{9}$

22. $\sqrt[3]{64}$

23. $\sqrt[3]{-27}$

24. $\sqrt[4]{1}$

25. $\sqrt[5]{-243}$

26. $\sqrt[6]{x^{18}}$

27. $\sqrt{x^6}$

28. $\sqrt[5]{y^{10}}$

29. $\sqrt[5]{-a^{15}}$

30. $\sqrt[5]{-32}$

31. $\sqrt[4]{r^8}$

32. $\sqrt[9]{-c^{18}}$

33. $\sqrt[5]{3125}$

34. $\sqrt[3]{c^{27}}$

35. $-\sqrt[5]{32}$

36. $\sqrt[3]{a^6}$

37. $-\sqrt[5]{-32}$

38. $-\sqrt[3]{8}$

39. $\sqrt[15]{-1}$

40. $\sqrt{49}$

41. $-\sqrt{144}$

42. $-\sqrt{\dfrac{49}{16}}$

43. $-\sqrt{\dfrac{121}{25}}$

44. $\sqrt{(-5)^2}$

45. $\sqrt{t^2}$

46. $\sqrt{s^{16}}$

47. $\sqrt{p^{20}}$

48. $\sqrt{a^{36}}$

Objective 4   **Graph functions defined by radical expressions.**

Graph each function and give its domain and its range.

49. $f(x) = \sqrt{x+1}$

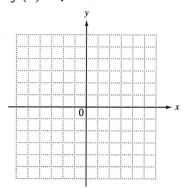

52. $f(x) = \sqrt{x} + 2$

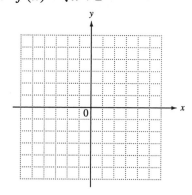

50. $f(x) = \sqrt{x-3}$

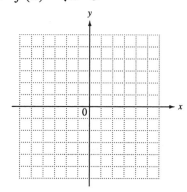

53. $f(x) = \sqrt[3]{x} - 2$

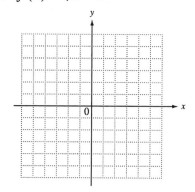

51. $f(x) = \sqrt{x} - 1$

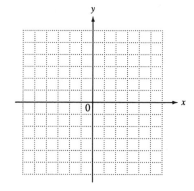

54. $f(x) = \sqrt[3]{x} + 2$

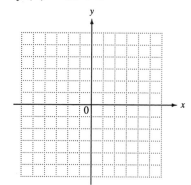

Objective 5   **Find nth roots of the nth powers.**

Simplify each root.

55. $\sqrt[4]{x^4}$

57. $\sqrt[5]{x^5}$

59. $\sqrt[6]{x^{12}}$

61. $-\sqrt[4]{x^{16}}$

56. $-\sqrt[4]{x^4}$

58. $-\sqrt[5]{x^5}$

60. $-\sqrt[5]{x^{10}}$

62. $\sqrt[7]{x^{21}}$

Objective 6   **Use a calculator to find roots.**

Use a calculator to find a decimal approximation for each radical. Round answers to three decimal places if necessary.

63. $-\sqrt{87}$

68. $\sqrt[4]{84}$

73. $-\sqrt{310}$

78. $-\sqrt[4]{130}$

64. $\sqrt{76}$

69. $\sqrt{101}$

74. $\sqrt[6]{200}$

79. $-\sqrt{370}$

65. $-\sqrt[3]{61}$

70. $-\sqrt{159}$

75. $\sqrt[3]{660}$

80. $-\sqrt[5]{990}$

66. $-\sqrt[3]{35}$

71. $\sqrt[5]{204}$

76. $\sqrt{870}$

81. $\sqrt{123}$

67. $\sqrt[4]{42}$

72. $\sqrt[3]{263}$

77. $\sqrt{930}$

82. $\sqrt[3]{28}$

9.1   Mixed Exercises

Find each root or use a calculator to find a decimal approximation. Round to three decimal places if necessary.

83. $\sqrt{196}$

86. $\sqrt{384}$

89. $\sqrt{k^4}$

92. $\sqrt[3]{43}$

84. $\sqrt[4]{4096}$

87. $-\sqrt{333}$

90. $\sqrt{d^2}$

93. $-\sqrt{-39}$

85. $\sqrt[3]{-3375}$

88. $\sqrt[4]{720}$

91. $\sqrt[5]{-p^{10}}$

94. $\sqrt[4]{45}$

9.2   Rational Exponents

In all exercises, assume that variables represent positive real numbers.

Objective 1   Use exponential notation for nth roots.

Evaluate each exponential.

1. $27^{1/3}$

2. $25^{1/2}$

3. $32^{1/5}$

4. $343^{1/3}$

5. $-81^{1/4}$

6. $(-81)^{1/4}$

7. $1000^{1/3}$

8. $(-8)^{1/3}$

9. $625^{1/4}$

10. $1024^{1/10}$

11. $-256^{1/4}$

12. $16^{1/4}$

13. $216^{1/3}$

14. $243^{1/5}$

15. $3375^{1/3}$

16. $512^{1/3}$

17. $81^{1/2}$

18. $125^{1/3}$

19. $1296^{1/4}$

20. $(-125)^{1/3}$

Objective 2   Define $a^{m/n}$.

Evaluate each exponential.

21. $27^{2/3}$

22. $25^{3/2}$

23. $81^{5/4}$

24. $27^{-2/3}$

25. $25^{-3/2}$

26. $243^{-2/5}$

27. $1000^{2/3}$

28. $-125^{-2/3}$

29. $16^{3/4}$

30. $216^{-2/3}$

31. $81^{-3/4}$

32. $729^{-1/6}$

33. $32^{3/5}$

34. $-8^{-2/3}$

35. $(-8)^{-2/3}$

36. $-625^{3/4}$

37. $512^{-2/3}$

38. $36^{5/2}$

39. $729^{5/6}$

40. $8^{-1/3}$

Objective 3   Convert between radicals and rational exponents.

Simplify each radical by rewriting it with a rational exponent. Write answers in radical form if necessary.

41. $\sqrt{x^{12}}$

42. $\sqrt[6]{4t^4}$

43. $\sqrt[3]{x^2} \cdot \sqrt[6]{x}$

44. $\sqrt{7^4}$

45. $\sqrt[8]{16x^{12}}$

46. $\sqrt[40]{y^{35}}$

47. $\sqrt[4]{5} \cdot \sqrt{5}$

48. $\sqrt[3]{2} \cdot \sqrt[4]{2}$

49. $\sqrt[3]{x} \cdot \sqrt{x}$

50. $\sqrt[5]{p^7}$

51. $\dfrac{\sqrt{r}}{\sqrt[3]{r}}$

52. $\sqrt[15]{27t^6}$

53. $\sqrt[3]{k^2} \cdot \sqrt[6]{k}$

54. $\sqrt[3]{p} \cdot \sqrt[5]{p^2}$

55. $\sqrt[3]{r^7}$

56. $\sqrt[4]{x^3} \cdot \sqrt[3]{x}$

57. $\sqrt[3]{x} \cdot \sqrt[4]{y} \cdot \sqrt{z}$

58. $\sqrt[4]{5} \cdot \sqrt{2}$

Objective 4 **Use the rules for exponents with rational exponents.**

Use the rules of exponents to simplify each expression. Write all answers with positive exponents.

59. $13^{4/5} \cdot 13^{6/5}$

60. $\dfrac{5^{3/7}}{5^{4/7}}$

61. $\dfrac{8^{3/4}}{8^{-1/4}}$

62. $(7^{2/3})^6$

63. $3^{1/2} \cdot 3^{3/2}$

64. $5^{3/4} \cdot 5^{9/4}$

65. $r^{2/3} \cdot r^{1/4}$

66. $y^{7/3} \cdot y^{-4/3}$

67. $\dfrac{a^{2/3} \cdot a^{-1/3}}{(a^{-1/6})^3}$

68. $(a^4)^{1/2} \cdot (a^6)^{1/3}$

69. $(5^{4/3})^6$

70. $(a^{-1})^{1/2}(a^{-3})^{-1/2}$

71. $x^{3/4} \cdot x^{5/6}$

72. $\dfrac{8^{3/5} \cdot 8^{-8/5}}{8^{-2}}$

73. $\dfrac{a^{4/5}}{a^{2/3}}$

74. $\left(\dfrac{c^6}{x^3}\right)^{2/3}$

75. $\dfrac{(x^{-3}y^2)^{2/3}}{(x^2y^{-5})^{2/5}}$

76. $\dfrac{(x^{-1}y^{2/3})^3}{(x^{1/3}y^{1/2})^2}$

9.2 Mixed Exercises

Simplify. Write answers with positive exponents only.

77. $-243^{1/5}$

78. $128^{1/7}$

79. $\left(-\dfrac{32}{3125}\right)^{3/5}$

80. $\left(\dfrac{81}{256}\right)^{-3/4}$

81. $-1000^{2/3}$

82. $-32^{-3/5}$

83. $625^{-3/4}$

84. $100,000^{2/5}$

85. $\dfrac{w^{7/4}w^{-1/2}}{w^{5/4}}$

86. $\left(\dfrac{x^{1/4}}{x^{-3/4}}\right)^2$

Simplify by first converting to rational exponents. Write answers in radical form if necessary.

87. $\sqrt[3]{x^2} \cdot \sqrt[9]{x^4}$

88. $\dfrac{\sqrt[5]{x^2}}{\sqrt[3]{x}}$

9.3   Simplifying Radical Expressions

In all exercises, assume that variables represent positive real numbers.

Objective 1   Use the product rule for radicals.

Multiply.

1. $\sqrt{3} \cdot \sqrt{11}$

2. $\sqrt{14} \cdot \sqrt{5}$

3. $\sqrt{2} \cdot \sqrt{t}$

4. $\sqrt{7x} \cdot \sqrt{6t}$

5. $\sqrt{\dfrac{7}{c}} \cdot \sqrt{\dfrac{13}{w}}$

6. $\sqrt{\dfrac{11}{r}} \cdot \sqrt{\dfrac{3}{p}}$

7. $\sqrt[4]{7} \cdot \sqrt[4]{6}$

8. $\sqrt[6]{4t} \cdot \sqrt[6]{5t^4}$

9. $\sqrt[5]{6r^2t^3} \cdot \sqrt[5]{4r^2t}$

10. $\sqrt[3]{3} \cdot \sqrt[3]{7}$

11. $\sqrt{5} \cdot \sqrt{7}$

12. $\sqrt{11} \cdot \sqrt{15}$

13. $\sqrt{6} \cdot \sqrt{r}$

14. $\sqrt[3]{5} \cdot \sqrt[3]{x}$

15. $\sqrt[3]{7x} \cdot \sqrt[3]{5y}$

16. $\sqrt[4]{8} \cdot \sqrt[4]{15}$

17. $\sqrt[7]{8a^2t^3} \cdot \sqrt[7]{6at^3}$

18. $\sqrt[4]{2} \cdot \sqrt[4]{2x}$

Objective 2   Use the quotient rule for radicals.

Simplify each radical.

19. $\sqrt{\dfrac{25}{16}}$

20. $\sqrt{\dfrac{5}{36}}$

21. $\sqrt{\dfrac{x}{81}}$

22. $\sqrt{\dfrac{t^9}{25}}$

23. $\sqrt[3]{-\dfrac{27}{8}}$

24. $\sqrt[3]{\dfrac{45}{27}}$

25. $\sqrt{\dfrac{49}{100}}$

26. $\sqrt{\dfrac{6}{49}}$

27. $\sqrt[3]{\dfrac{125}{r^{15}}}$

28. $\sqrt[3]{-\dfrac{a^6}{125}}$

29. $\sqrt[3]{\dfrac{w^3}{216}}$

30. $\sqrt[3]{\dfrac{\ell^2}{27}}$

31. $\sqrt{\dfrac{r}{121}}$

32. $\sqrt[4]{\dfrac{p}{16}}$

33. $\sqrt[3]{-\dfrac{343}{125}}$

34. $\sqrt{\dfrac{z^4}{36}}$

35. $\sqrt[5]{\dfrac{7x}{32}}$

36. $\sqrt{\dfrac{15}{169}}$

Objective 3   Simplify radicals.

Express each radical in simplified form.

37. $\sqrt{27}$

38. $\sqrt{63}$

39. $\sqrt{75}$

40. $\sqrt{200}$

41. $\sqrt{135}$

42. $\sqrt[3]{24}$

43. $\sqrt[4]{32}$

44. $\sqrt{36t^5}$

45. $\sqrt{50r^5x}$

46. $\sqrt[3]{24y^6}$

47. $\sqrt[3]{54x^{11}}$

48. $\sqrt[3]{80x^9c^7}$

49. $\sqrt[3]{108a^5}$ **50.** $\sqrt{49r^9}$ **51.** $\sqrt[3]{270b^4c^8}$

Objective 4   **Simplify products and quotients of radicals with different indexes.**

Express each radical in simplified form.

52. $\sqrt[12]{x^{16}}$ **55.** $\sqrt[10]{x^{25}}$ **58.** $\sqrt[24]{z^{30}}$ **61.** $\sqrt[12]{9^3 x^6 y^9}$

53. $\sqrt[6]{x^9}$ **56.** $\sqrt[9]{9^3}$ **59.** $\sqrt[20]{x^{15} y^{10}}$ **62.** $\sqrt{6} \cdot \sqrt[3]{5}$

54. $\sqrt[24]{5^4}$ **57.** $\sqrt[42]{x^{28}}$ **60.** $\sqrt[6]{z^4 y^2}$ **63.** $\sqrt[3]{3} \cdot \sqrt[6]{7}$

Objective 5   **Use the Pythagorean formula.**

Find the missing length in each right triangle. Simplify the answer if necessary.

64.

68.

72.

65.

69.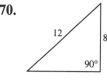

73.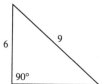

66.

67.

70.

71.

Objective 6   **Use the distance formula.**

Find the distance between each pair of points.

74. $(2, 5)$ and $(-2, 3)$

75. $(3, 4)$ and $(-1, -2)$

76. $(1, -2)$ and $(4, -3)$

77. $(-1, 3)$ and $(5, -2)$

78. $(-1, -2)$ and $(-4, 3)$

79. $(4, 7)$ and $(-3, 8)$

80. $(6, -5)$ and $(9, 2)$

81. $(-4, 7)$ and $(5, -3)$

82. $\left(\sqrt{5}, -\sqrt{2}\right)$ and $\left(2\sqrt{5}, 3\sqrt{2}\right)$

83. $\left(4\sqrt{3}, 2\sqrt{5}\right)$ and $\left(3\sqrt{3}, -\sqrt{5}\right)$

84. $(2x, x + y)$ and $(x, x - y)$

85. $(a - b, a)$ and $(a + b, b)$

9.3   Mixed Exercises

Simplify.

86. $\sqrt{5a} \cdot \sqrt{6b}$

87. $\sqrt[4]{16x^{12}y^{10}}$

88. $\sqrt[5]{4w} \cdot \sqrt[5]{2w^3}$

89. $\sqrt[3]{54t^7}$

90. $\sqrt[3]{125}$

91. $\sqrt{3x^7y}$

92. $\sqrt[3]{\dfrac{27}{8}}$

93. $\sqrt{\dfrac{a^4}{625}}$

94. $\sqrt[3]{\dfrac{b^{13}q^6}{8}}$

95. $\sqrt{8x^3y^6z^{11}}$

96. $\sqrt[4]{t^3v^{12}}$

97. $\sqrt{3} \cdot \sqrt[5]{64}$

Find the missing length in the right triangle. Simplify the answer if necessary.

98.

99.

Find the distance between each pair of points.

100. $(-1, 6)$ and $(7, 4)$

101. $(5, -1)$ and $(2, -5)$

9.4   Adding and Subtracting Radical Expressions

Objective 1   Simplify radical expressions involving addition and subtraction.

Add or subtract. Assume that all variables represent positive real numbers.

1. $2\sqrt{7}+3\sqrt{7}$

2. $3\sqrt{13}+5\sqrt{52}$

3. $3\sqrt{48}+5\sqrt{27}$

4. $2\sqrt{18}-5\sqrt{32}+7\sqrt{162}$

5. $5\sqrt[3]{24}+4\sqrt[3]{81}$

6. $7\sqrt[4]{32}-9\sqrt[4]{2}$

7. $5\sqrt{13}+4\sqrt{13}-6\sqrt{13}$

8. $3\sqrt{6}-8\sqrt{6}-5\sqrt{24}$

9. $\sqrt{48y}+\sqrt{12y}+\sqrt{27y}$

10. $\sqrt{98}-2\sqrt{8}+\sqrt{32}$

11. $\sqrt[3]{81}+\sqrt[3]{24}+\sqrt[3]{192}$

12. $\sqrt[3]{625}+\sqrt[3]{135}-\sqrt[3]{40}$

13. $\sqrt{100x}-\sqrt{9x}+\sqrt{25x}$

14. $-2\sqrt[3]{81}+\sqrt[3]{24}$

Solve each problem. Give answers as simplified radical expressions.

15. Find the perimeter of a square with a side that measures $\sqrt{18}$ in. Find the area of the square.

16. Find the perimeter of a triangle with sides a, b, and c if $a=\sqrt{20}$ cm, $b=\sqrt{80}$ cm, and $c=\sqrt{125}$ cm.

17. Find the perimeter of a rectangle whose length is $\sqrt{12}$ ft and whose width is $\sqrt{27}$ ft. What is the area of the rectangle?

18. Use $A=\frac{1}{2}h(B+b)$ to find A if $h=10\sqrt{5}$ cm, $B=6\sqrt{3}$ cm, and $b=\sqrt{3}$ cm.

19. Use $A=\frac{1}{2}h(B+b)$ to find A if $b=2\sqrt{5}$ m, $h=3\sqrt{2}$ m, and $B=4\sqrt{5}$ m.

9.4   Mixed Exercises

Add or subtract. Assume that all variables represent positive real numbers.

20. $5\sqrt{10}-\sqrt{10}-3\sqrt{10}$

21. $3\sqrt{54}-5\sqrt{24}$

22. $7\sqrt[3]{54}-6\sqrt[3]{128}$

23. $6\sqrt[3]{135}+3\sqrt[3]{40}$

24. $2\sqrt[3]{16r}+\sqrt[3]{54r}-\sqrt[3]{16r}$

25. $3\sqrt{8z}+3\sqrt{2z}+\sqrt{32z}$

26. $3\sqrt{18z}+2\sqrt{8z}$

27. $6\sqrt[3]{54z^3}+5\sqrt[3]{16z^3}$

Solve each problem. Give answers as simplified radical expressions.

28. Find the width of a rectangle with length $3\sqrt{75}$ cm and perimeter $36\sqrt{3}$ cm. Find the area of the rectangle.

29. Use $A = \frac{1}{2}bh$ to find b if $A = 6\sqrt{105}$ in.^{2} and $h = 6\sqrt{3}$ in.

9.5   Multiplying and Dividing Radical Expressions

In all exercises, assume that variables represent positive real numbers.

Objective 1   Multiply radical expressions

Multiply each product, then simplify.

1. $\left(3+\sqrt{2}\right)\left(2+\sqrt{7}\right)$

2. $\left(\sqrt{10}+\sqrt{3}\right)\left(\sqrt{6}-\sqrt{11}\right)$

3. $\left(5+\sqrt{2}\right)\left(5-\sqrt{2}\right)$

4. $\left(\sqrt{5}+\sqrt{6}\right)\left(\sqrt{2}-4\right)$

5. $\left(\sqrt{2}+4\right)\left(\sqrt{2}-4\right)$

6. $\left(3\sqrt{3}+5\right)^2$

7. $\left(3\sqrt{2}-4\right)\left(2\sqrt{2}+7\right)$

8. $\left(\sqrt{2}-\sqrt{12}\right)^2$

9. $\left(\sqrt{5}+1\right)\left(\sqrt{5}-1\right)$

10. $\left(\sqrt{x}+y\right)\left(\sqrt{x}-y\right)$

11. $\left(2\sqrt{x}-3\right)\left(3\sqrt{x}-2\right)$

12. $\left(2+\sqrt[3]{5}\right)\left(2-\sqrt[3]{5}\right)$

Objective 2   Rationalize denominators with one radical term.

Rationalize the denominator in each expression.

13. $\dfrac{6}{\sqrt{5}}$

14. $\dfrac{15}{\sqrt{5}}$

15. $\dfrac{\sqrt{2}}{\sqrt{11}}$

16. $\dfrac{\sqrt{15}}{\sqrt{2}}$

17. $\dfrac{3}{\sqrt{18}}$

18. $\dfrac{3}{\sqrt{8}}$

19. $\dfrac{7}{\sqrt{75}}$

20. $-\dfrac{14}{\sqrt{27}}$

21. $\dfrac{12}{\sqrt{50}}$

22. $\dfrac{4}{\sqrt{6}}$

23. $\dfrac{7}{4\sqrt{3}}$

24. $\dfrac{\sqrt{3}}{8\sqrt{5}}$

25. $\dfrac{5}{2\sqrt{5}}$

26. $\dfrac{5}{\sqrt{10}}$

27. $\dfrac{3}{\sqrt{98}}$

Simplify.

28. $\sqrt{\dfrac{27}{48}}$

29. $\sqrt{\dfrac{36}{t}}$

30. $\sqrt{\dfrac{50}{r}}$

31. $\sqrt{\dfrac{162x^4}{t^5}}$

32. $\sqrt{\dfrac{8}{m}}$

33. $\sqrt{\dfrac{25}{y}}$

34. $\sqrt{\dfrac{5x}{8}}$

35. $\sqrt{\dfrac{4x^2}{3}}$

36. $\sqrt{\dfrac{18z}{7}}$

37. $\sqrt{\dfrac{4}{7y}}$

43. $\sqrt[3]{\dfrac{1}{9}}$

49. $\sqrt[3]{\dfrac{7}{36}}$

38. $\sqrt{\dfrac{5}{8x^2}}$

44. $\sqrt[3]{\dfrac{14}{243}}$

50. $\sqrt[3]{\dfrac{4}{5}}$

39. $\sqrt{\dfrac{12a^2}{5t^3}}$

45. $\sqrt[3]{\dfrac{8}{100}}$

51. $\sqrt[3]{\dfrac{c^3}{d^2}}$

40. $\sqrt{\dfrac{7y^2}{12b}}$

46. $\sqrt[3]{\dfrac{x}{y}}$

52. $\sqrt[3]{\dfrac{t^6}{x^7}}$

41. $\sqrt{\dfrac{19}{32}}$

47. $\sqrt[3]{\dfrac{3}{8}}$

53. $\sqrt[3]{\dfrac{r}{98}}$

42. $\sqrt{\dfrac{1}{5}}$

48. $\sqrt[3]{\dfrac{5x}{2}}$

54. $\sqrt[3]{\dfrac{m}{2x}}$

Objective 3 **Rationalize denominators with binomials involving radicals.**

Rationalize each denominator.

55. $\dfrac{5}{7-\sqrt{3}}$

60. $-\dfrac{4}{\sqrt{7}-\sqrt{5}}$

65. $\dfrac{7}{\sqrt{3}-1}$

56. $\dfrac{-6}{\sqrt{7}+3}$

61. $\dfrac{5}{\sqrt{3}+\sqrt{11}}$

66. $-\dfrac{6}{\sqrt{6}-\sqrt{3}}$

57. $\dfrac{5}{3-\sqrt{5}}$

62. $\dfrac{1}{4+\sqrt{5}}$

67. $-\dfrac{5}{\sqrt{5}-\sqrt{3}}$

58. $\dfrac{12}{6+\sqrt{8}}$

63. $\dfrac{1}{4-\sqrt{7}}$

68. $\dfrac{\sqrt{3}}{\sqrt{5}-\sqrt{2}}$

59. $\dfrac{18}{\sqrt{11}+\sqrt{2}}$

64. $\dfrac{2}{\sqrt{2}-\sqrt{3}}$

69. $\dfrac{\sqrt{6}}{\sqrt{13}+\sqrt{5}}$

Objective 4 **Write radical quotients in lowest terms.**

Write each quotient in lowest terms.

70. $\dfrac{12-3\sqrt{2}}{3}$

72. $\dfrac{9+6\sqrt{15}}{12}$

74. $\dfrac{7-\sqrt{98}}{14}$

71. $\dfrac{30-45\sqrt{3}}{20}$

73. $\dfrac{35-7\sqrt{6}}{7}$

75. $\dfrac{5+2\sqrt{75}}{25}$

76. $\dfrac{12+18\sqrt{3}}{6}$

78. $\dfrac{8-2\sqrt{12}}{4}$

80. $\dfrac{6+2\sqrt{5}}{2}$

77. $\dfrac{2x-\sqrt{8x^2}}{4x}$

79. $\dfrac{50+\sqrt{80x}}{10}$

81. $\dfrac{16-12\sqrt{72}}{24}$

9.5   Mixed Exercises

Multiply, then simplify each product.

82. $\sqrt{3}\left(\sqrt{18}-2\sqrt{12}\right)$

83. $\left(\sqrt[3]{2}+1\right)\left(\sqrt[3]{2}-1\right)$

Rationalize the denominator in each expression.

84. $-\dfrac{15}{\sqrt{8}}$

86. $\dfrac{3\sqrt{2}}{\sqrt{t}}$

88. $\dfrac{3\sqrt{x}}{2\sqrt{t}+\sqrt{x}}$

85. $\dfrac{12}{\sqrt{27}}$

87. $\dfrac{3-\sqrt{7}}{\sqrt{3}-7}$

89. $\dfrac{\sqrt{t}}{\sqrt{t}+3}$

Simplify.

90. $\sqrt{\dfrac{27}{t^3}}$

93. $\sqrt{\dfrac{5a}{3b}}$

96. $\sqrt[3]{\dfrac{s^2}{6}}$

91. $\sqrt{\dfrac{3}{10}}$

94. $\sqrt{\dfrac{x^2}{y^3}}$

97. $\sqrt[3]{\dfrac{t^{15}}{x}}$

92. $\sqrt{\dfrac{c}{x^2}}$

95. $\sqrt[3]{\dfrac{a^4}{b}}$

98. $\sqrt[3]{\dfrac{2x}{3z}}$

Write each quotient in lowest terms.

99. $\dfrac{12-3\sqrt{8}}{9}$

100. $\dfrac{7y-\sqrt{98y^5}}{14y}$

101. $\dfrac{12+\sqrt{20}}{10}$

102. $\dfrac{5+10\sqrt{2}}{5}$

103. $\dfrac{3+\sqrt{18x}}{3}$

104. $\dfrac{5\sqrt{5}-\sqrt{25}}{35}$

9.6   Solving Equations with Radicals

Objective 1   Solve radical equations using the power rule.

Solve each equation.

1. $\sqrt{t} = 5$

2. $\sqrt{4x+1} = 3$

3. $\sqrt{3t+7} = 5$

4. $\sqrt{9c+9} = 9$

5. $\sqrt{7x-6} = 8$

6. $\sqrt{6x+1} = 1$

7. $\sqrt{2m+6} = 6$

8. $\sqrt{r+8} = 3$

9. $\sqrt{2p+5} = 5$

10. $\sqrt{4x-19} = 5$

11. $\sqrt{5p-5} = 5$

12. $\sqrt{x-7} = 3$

13. $\sqrt{2q-1} = 9$

14. $\sqrt{12h+1} = 7$

15. $\sqrt{6x+25} = 25$

16. $\sqrt{3w+4} = 7$

17. $\sqrt{9c+1} = 8$

18. $\sqrt{3x+1} = 8$

19. $\sqrt{x+2} = 3$

20. $\sqrt{12x+16} = 16$

21. $\sqrt{t} = -6$

22. $\sqrt{a} - 6 = -2$

23. $\sqrt{x-7} - 4 = 0$

24. $\sqrt{c+3} + 7 = 0$

25. $\sqrt{t+1} - 4 = 0$

26. $\sqrt{2r+6} - 4 = 0$

27. $\sqrt{12p+1} + 7 = 0$

28. $\sqrt{y-4} - 3 = 0$

29. $\sqrt{2x+5} - 3 = 0$

30. $\sqrt{5x-4} - 1 = 0$

31. $\sqrt{x+3} - 5 = 0$

32. $\sqrt{2x+3} - 5 = 0$

33. $\sqrt{7y-5} - 3 = 0$

34. $\sqrt{5r-4} - 9 = 0$

35. $\sqrt{m} + 7 = -1$

36. $\sqrt{x+13} - 5 = 2$

Objective 2   Solve radical equations that require additional steps.

Solve each equation.

37. $\sqrt{6a-23} = 3a-11$

38. $\sqrt{2x+7} = 2x+1$

39. $\sqrt{4x+17} = x+3$

40. $\sqrt{x-3} = x-3$

41. $\sqrt{x-1} = x-7$

42. $\sqrt{13-3x} = x-5$

43. $\sqrt{8x+33} = x+3$

44. $\sqrt{41-16x} = x-6$

45. $\sqrt{30-5x} = x-6$

46. $\sqrt{7x+39} = x+3$

47. $\sqrt{7y+15} = 2y+3$

48. $\sqrt{3q-8} = q-2$

49. $\sqrt{33-8r} = 2r-3$

50. $\sqrt{44-20x} = -8x$

51. $\sqrt{25-8x} = x-2$

52. $\sqrt{12x+4} = x+2$

53. $\sqrt{58-11w} = w+2$

54. $\sqrt{27-18v} = 2v-3$

55. $\sqrt{t+2} - \sqrt{t-3} = 1$

56. $\sqrt{c+1} + \sqrt{4c-3} = 5$

57. $\sqrt{3k+7} + \sqrt{k+1} = 2$

58. $\sqrt{x+1}+\sqrt{7x+4}=7$

59. $\sqrt{2p+5}+\sqrt{p+51}=8$

60. $\sqrt{4w-3}+\sqrt{w+2}=8$

61. $\sqrt{5y+4}-\sqrt{2y+2}=1$

62. $\sqrt{x+4}+\sqrt{9-4x}=5$

63. $\sqrt{x-4}+\sqrt{x+11}=5$

64. $\sqrt{3x+3}-\sqrt{2x+3}=1$

65. $\sqrt{5x+9}-\sqrt{3x+4}=1$

66. $\sqrt{2-r}+\sqrt{r+11}=5$

67. $\sqrt{5c+6}-\sqrt{c+3}=3$

68. $\sqrt{x+12}+\sqrt{14-c}=6$

69. $\sqrt{7x+4}-\sqrt{x+6}=2$

70. $\sqrt{11x+22}-\sqrt{9x+19}=1$

71. $\sqrt{3p+10}+\sqrt{p+2}=6$

72. $\sqrt{k+10}+\sqrt{2k+19}=2$

Objective 3 Solve radical equations with indexes greater than 2.

Solve each equation.

73. $\sqrt[3]{7x-5}-\sqrt[3]{3x+7}=0$

74. $\sqrt[4]{3x-3}=3$

75. $\sqrt[3]{t^2+5t+15}=\sqrt[3]{t^2}$

76. $\sqrt[4]{3k+2}+2=0$

77. $\sqrt[5]{2t+1}-1=0$

78. $\sqrt[4]{8x+5}=\sqrt[4]{7x+7}$

79. $\sqrt[3]{5r-6}-\sqrt[3]{3r+4}=0$

80. $\sqrt[5]{2w+5}=\sqrt[5]{7w}$

81. $\sqrt[3]{4x-4}=\sqrt[3]{3x+13}$

82. $\sqrt[4]{c^2+2c+18}=\sqrt[4]{c^2}$

83. $\sqrt[5]{b-1}-2=0$

84. $\sqrt[3]{5x-12}=\sqrt[3]{6x-15}$

85. $\sqrt[5]{5a+1}-\sqrt[5]{2a-11}=0$

86. $\sqrt[3]{6x-4}=\sqrt[3]{2x+8}$

87. $\sqrt[3]{2a-63}+5=0$

88. $\sqrt[3]{3x-11}=\sqrt[3]{2x+10}$

89. $\sqrt[3]{8-x}-2=0$

90. $\sqrt[3]{x^3-125}+5=0$

9.6 Mixed Exercises

Solve each equation.

91. $\sqrt{3q-1}+5=0$

92. $\sqrt{x}-8=0$

93. $\sqrt{28x-6}=2x+3$

94. $\sqrt[3]{4t-1}-3=0$

95. $\sqrt{2w+25}+\sqrt{2w+16}=9$

96. $\sqrt{7r+8}-\sqrt{r+1}=5$

97. $\sqrt[3]{r^2+3r+15}-\sqrt[3]{r^2}=0$

98. $\sqrt{h+7}=2$

99. $\sqrt{30t+19}=2t+5$

100. $\sqrt{x+12}-4=0$

9.7   Complex Numbers

Objective 1 Simplify numbers of the form $\sqrt{-b}$, where $b > 0$.

Write each number as a product of a real number and i. Simplify all radical expressions.

1. $\sqrt{-49}$

6. $\sqrt{-50}$

11. $\sqrt{-60}$

16. $\sqrt{-1080}$

2. $\sqrt{-36}$

7. $-\sqrt{-63}$

12. $\sqrt{-450}$

17. $-\sqrt{-625}$

3. $-\sqrt{-100}$

8. $\sqrt{-120}$

13. $-\sqrt{-125}$

18. $-\sqrt{-162}$

4. $\sqrt{-6}$

9. $\sqrt{-18}$

14. $-\sqrt{-72}$

5. $\sqrt{-22}$

10. $-\sqrt{-27}$

15. $\sqrt{-99}$

Multiply or divide as indicated.

19. $\sqrt{-6} \cdot \sqrt{-6}$

20. $\sqrt{-3} \cdot \sqrt{-15}$

21. $\sqrt{-21} \cdot \sqrt{-7}$

22. $\sqrt{-14} \cdot \sqrt{3}$

23. $\sqrt{-3} \cdot \sqrt{7}$

24. $\sqrt{2} \cdot \sqrt{-5} \cdot \sqrt{3}$

25. $\sqrt{-6} \cdot \sqrt{-3} \cdot \sqrt{2}$

26. $\sqrt{-5} \cdot \sqrt{-3} \cdot \sqrt{-7}$

27. $\dfrac{\sqrt{-125}}{\sqrt{-5}}$

28. $\dfrac{\sqrt{-16}}{\sqrt{2}}$

29. $\dfrac{\sqrt{-28}}{\sqrt{7}}$

30. $\dfrac{\sqrt{-18}}{\sqrt{-2}}$

31. $\dfrac{\sqrt{-45}}{\sqrt{-5}}$

32. $\dfrac{\sqrt{-80}}{\sqrt{5}}$

33. $\dfrac{\sqrt{-200}}{\sqrt{-8}}$

34. $\dfrac{\sqrt{-42} \cdot \sqrt{-6}}{\sqrt{-7}}$

35. $\dfrac{\sqrt{-56} \cdot \sqrt{-6}}{\sqrt{16}}$

36. $\dfrac{\sqrt{-10} \cdot \sqrt{7}}{\sqrt{25}}$

Objective 2 Recognize subsets of the complex numbers.

*The real numbers are a subset of the complex numbers. Classify each of the following complex numbers as **real** or **imaginary**.*

37. $7 - 3i$

38. $-i\sqrt{10}$

39. $\sqrt{5}$

40. $\sqrt{3} - i\sqrt{5}$

41. $-\dfrac{1}{3}$

42. $-\dfrac{2}{3}i$

43. $\dfrac{3}{4} + \dfrac{5}{4}i$

44. $\sqrt{3} + \sqrt{5}$

45. $5 - i\sqrt{2}$

46. $i\sqrt{7}$

Objective 3 Add and subtract complex numbers.

Add or subtract as indicated. Write your answers in standard form.

47. $(5 + 7i) + (-2 + 4i)$

48. $(-2 + 9i) + (10 - 3i)$

49. $(5-2i)+(9+0i)$

50. $4+(3+6i)$

51. $(8+3i)-(5+3i)$

52. $(-2+3i)-(5+i)$

53. $(-7-2i)-(-3-3i)$

54. $7-(2-3i)$

55. $(7-3i)+(2-6i)$

56. $4i-(9+5i)+(2+3i)$

57. $(3+3i)-(8+4i)$

58. $(-1-5i)+(2+5i)$

59. $(7-9i)-(5-6i)$

60. $(5+3i)+(5-3i)$

61. $\left(\sqrt{3}-2i\sqrt{2}\right)+\left(2\sqrt{3}-2i\sqrt{2}\right)$

62. $[(8+4i)-(5-3i)]+(4-2i)$

Objective 4   **Multiply complex numbers.**

Multiply.

63. $(2+5i)(3-i)$

64. $(2-3i)(2+7i)$

65. $(4-6i)(2+6i)$

66. $(5-3i)(5+3i)$

67. $(7+2i)(7-2i)$

68. $(3-2i)(2-3i)$

69. $(1+3i)^2$

70. $\left(\sqrt{2}-i\sqrt{3}\right)^2$

71. $(8+2i)(3-5i)$

72. $(2+i)(3-i)$

Objective 5   **Divide complex numbers.**

Write each quotient in the form a + bi.

73. $\dfrac{1+i}{2-i}$

74. $\dfrac{5+2i}{9-4i}$

75. $\dfrac{7+9i}{4-i}$

76. $\dfrac{3-i}{i}$

77. $\dfrac{4+i}{3-2i}$

78. $\dfrac{3-2i}{2+i}$

79. $\dfrac{6-i}{2-3i}$

80. $\dfrac{3i}{2-i}$

81. $\dfrac{2-i}{4i}$

82. $\dfrac{1+i}{(2-i)(2+i)}$

Objective 6   **Find powers of *i*.**

Find each power of i.

83. i^{11}

84. i^{17}

85. i^{48}

86. i^{42}

87. i^{-13}

88. i^{-7}

89. i^{100}

90. i^{62}

91. i^{307}

92. i^{236}

93. i^{-100} **94.** i^{75} **95.** i^{-23} **96.** i^{115} **97.** i^{-21}

9.7 Mixed Exercises

Simplify.

98. $\sqrt{-125}$ **100.** i^9

99. $\sqrt{-4}$ **101.** i^{-3}

Perform the indicated operations. Write answers in the form a + bi.

102. $\sqrt{-7} \cdot \sqrt{-35}$ **108.** $(3+2i)(5-i)$

103. $\sqrt{-8} \cdot \sqrt{-8}$ **109.** $(2-5i)(2+5i)$

104. $(8-5i)+(3+i)$ **110.** $\dfrac{6+i}{2-3i}$

105. $(12+2i)-(-1+i)$

106. $(12+2i)(-1+i)$ **111.** $\dfrac{5-9i}{4-3i}$

107. $5+(-3+4i)$

Chapter 10

QUADRATIC EQUATIONS, INEQUALITIES, AND FUNCTIONS

10.1 Solving Quadratic Equations by the Square Root Property

Objective 1 Solve quadratic equations of the form $x^2 = k$, where **k > 0.**

Solve each equation by using the square root property. Express all radicals in simplest form.

1. $x^2 = 49$

2. $y^2 = 121$

3. $r^2 = 900$

4. $s^2 = 81$

5. $a^2 = 24$

6. $b^2 = -4$

7. $k^2 = 45$

8. $q^2 = 50$

9. $x^2 = 72$

10. $w^2 = 98$

11. $c^2 + 36 = 0$

12. $d^2 - 250 = 0$

13. $x^2 = \dfrac{9}{25}$

14. $z^2 = \dfrac{81}{64}$

15. $h^2 = \dfrac{90}{289}$

16. $x^2 = -\dfrac{36}{16}$

17. $m^2 = 1.96$

18. $t^2 - 12.25 = 0$

Objective 2 Solve quadratic equations of the form $(ax + b)^2 = k$, where **k > 0.**

Solve each equation by using the square root property. Express all radicals in simplest form.

19. $(x + 4)^2 = 25$

20. $(x - 3)^2 = 81$

21. $(y + 2)^2 = 16$

22. $(y - 9)^2 = 49$

23. $(x - 7)^2 = 0$

24. $(y + 8)^2 = 0$

25. $(m + 7)^2 = 4$

26. $(n + 3)^2 = 18$

27. $(p - 9)^2 = 28$

28. $(q - 2)^2 = 27$

29. $(r + 7)^2 = -25$

30. $(2x + 5)^2 = 32$

31. $(3m + 2)^2 = 27$

32. $(3p - 2)^2 - 28 = 0$

33. $(2p + 9)^2 - 4 = 0$

34. $(7p - 4)^2 = -81$

35. $\left(\dfrac{1}{9}x + 4\right)^2 = 16$

36. $\left(\dfrac{1}{5}x - 2\right)^2 = 49$

Objective 3 Solve quadratic equations with nonreal complex solutions.

Find the imaginary solutions of each equation.

37. $x^2 + 1 = 0$

38. $9 + y^2 = 0$

39. $p^2 = -16$

40. $9 = -(6q - 7)^2$

41. $k^2 + 25 = 0$

42. $(m + 1)^2 = -36$

43. $4x^2 + 25 = 0$

44. $121 + q^2 = 0$

46. $49 + 16(2w+4)^2 = 0$

45. $-81 - k^2 = 0$

10.1 Mixed Exercises

Solve each equation.

47. $(a+5)^2 = 7$

48. $b^2 - 225 = 0$

49. $z^2 = 27$

50. $(2s+4)^2 = 7$

51. $(x-3)^2 = -1$

52. $(4a+5)^2 = -12$

53. $\dfrac{9x^2}{2} = 50$

54. $2a^2 = 20$

55. $x^2 + 3x = 0$

56. $x^2 + 7x + 1 = 0$

57. $2a^2 + 7a - 13 = 0$

58. $x^2 - 2x + 3 = 0$

59. $r^2 - 5r + 4 = 0$

60. $x^2 - 3 = \dfrac{1}{2}x$

10.2 Solving Quadratic Equations by Completing the Square

Objective 1 Solve quadratic equations by completing the square when the coefficient of the squared term is 1.

Solve each equation by completing the square.

1. $x^2 - 2x = 15$

2. $x^2 - 5x = 14$

3. $p^2 + 6p = 0$

4. $x^2 + 3x = 4$

5. $m^2 - 6m = -12$

6. $r^2 + 8r = -4$

7. $x^2 - 10x = -21$

8. $x^2 - 4x = 2$

9. $x^2 + 4x - 2 = 0$

10. $w^2 - 3w - 4 = 0$

11. $b^2 + 5b - 5 = 0$

12. $x^2 + 2x = 63$

13. $x^2 - 8x + 16 = 0$

14. $d^2 + 10d - 11 = 0$

15. $z^2 + 3z - \dfrac{7}{4} = 0$

16. $c^2 - c - \dfrac{5}{2} = 0$

Objective 2 Solve quadratic equations by completing the square when the coefficient of the squared term is not 1.

Solve each equation by completing the square.

17. $4x^2 + 8x = 0$

18. $3m^2 - 15m = 42$

19. $2x^2 - 13x + 20 = 0$

20. $3r^2 = 6r + 2$

21. $3z^2 + 3z - 4 = 0$

22. $2p^2 + 6p - 1 = 0$

23. $6a^2 - 4a = -5$

24. $6x^2 - x = 15$

25. $2x^2 + 13x - 7 = 0$

26. $3n^2 + 7n + 2 = 0$

27. $-r^2 + 3r = -1$

28. $-y^2 - 2y + 8 = 0$

29. $3r^2 - 6r - 2 = 0$

30. $3t^2 + t - 2 = 0$

31. $3x^2 - 2x + 4 = 0$

32. $6q^2 + 4q = 1$

Objective 3 Simplify an equation before solving.

Simplify each of the following equations and then solve by completing the square.

33. $3p^2 = 3p + 5$

34. $2x - 4 = x^2 - 2x$

35. $4y^2 + 6y = 2y + 3$

36. $6y^2 + 3y = 4y^2 + y - 5$

37. $(x-2)(x+3) = 10$

38. $(b-1)(b+7) = 9$

39. $(c+3)(c+7) = 5$

40. $(s+3)(s+1) = 1$

41. $(x-1)(x+2) = 4$

42. $(j+3)(j-2) = 5$

10.2 Mixed Exercises

Solve each equation by completing the square.

43. $x^2 + 3x = 0$

44. $x^2 - 3x - 10 = 0$

45. $2m^2 + 6m = -4$

46. $4y^2 + 4y + 5 = 0$

47. $x^2 - x - 6 = 0$

48. $-x^2 - 4 + 2x = 0$

49. $2z^2 = 6z + 3 - 4z^2$

50. $6p - 3 = p^2 + 2p$

51. $-x^2 + 6x = 6$

52. $m^2 - 4m + 8 = 6m$

53. $(z + 7)(z - 3) = 4$

54. $3x^2 - 4x - 3 = 0$

10.3 Solving Quadratic Equations by the Quadratic Formula

Objective 1   Derive the quadratic formula.

Objective 2   Solve quadratic equations using the quadratic formula.

Use the quadratic formula to solve each equation.

1. $x^2 - 7x + 6 = 0$

2. $x^2 - 12x + 27 = 0$

3. $x^2 + 5x - 14 = 0$

4. $p^2 - 3p - 40 = 0$

5. $r^2 - 8r + 16 = 0$

6. $5t^2 - 13t + 6 = 0$

7. $6m^2 - 17m + 12 = 0$

8. $3w^2 + 6w - 24 = 0$

9. $16x^2 - 9 = 0$

10. $(z + 2)^2 = 2(5z - 2)$

11. $5m^2 + 5m - 1 = 0$

12. $4p(p + 1) = 1$

13. $3x^2 + 1 = 6x$

14. $2x^2 = 4x - 3$

15. $\dfrac{m}{2m - 1} = \dfrac{2m + 3}{15}$

16. $x^2 + 1 = \dfrac{13x}{6}$

Objective 3   Use the discriminant to determine the number and type of solutions.

Use the discriminant to determine whether the solutions for each equation are

A. *two rational numbers,*
C. *two irrational numbers,*

B. *one rational number,*
D. *two imaginary numbers.*

Do not actually solve.

17. $5a^2 - 4a + 1 = 0$

18. $2k^2 + 2k + 3 = 0$

19. $p^2 - 2p + 4 = 0$

20. $t^2 + 5t + 4 = 0$

21. $2y^2 + 4y + 8 = 0$

22. $3r^2 + 5r + 1 = 0$

23. $m^2 - 4m + 4 = 0$

24. $5y^2 - 5y + 2 = 0$

25. $4t^2 + 12t + 9 = 0$

26. $z^2 + 6z + 3 = 0$

27. $2r^2 = 4r + 3$

28. $3p^2 = 2p + 5$

29. $2m^2 - 4m = 8$

30. $n^2 + n = 2$

10.3 Mixed Exercises

Use the quadratic formula to solve each equation.

31. $t^2 - 2t + 3 = 0$

32. $x^2 - x - 3 = 0$

33. $r^2 = 12r + 28$

34. $4x^2 = -4x + 8$

Use the discriminant to determine whether the solutions for each equation are

A. *two rational numbers,*   **B.** *one rational number,*
C. *two irrational numbers,*   **D.** *two imaginary numbers.*

Do not actually solve.

35. $12r^2 - 40r + 25$

37. $5p^2 - 7p - 12$

39. $16x^2 - 40x + 25$

36. $6x^2 - 3x - 10$

38. $18p^2 + 46p - 24$

40. $16x^2 - 12x + 9$

10.4 Equations Quadratic in Form

Objective 1 Solve an equation with fractions by writing it in quadratic form.

Solve each equation. Check your solutions.

1. $1 + \dfrac{1}{x} = \dfrac{6}{x^2}$

2. $\dfrac{x}{2x+15} = \dfrac{1}{3x-2}$

3. $5 + \dfrac{6}{m+1} = \dfrac{14}{m}$

4. $\dfrac{7}{x^2} + 6 = -\dfrac{23}{x}$

5. $2 + \dfrac{9}{x} = \dfrac{2}{x+1}$

6. $4 - \dfrac{8}{x-1} = -\dfrac{35}{x}$

7. $2 = \dfrac{1}{x} + \dfrac{28}{x^2}$

8. $9 - \dfrac{12}{x} = -\dfrac{4}{x^2}$

9. $1 - \dfrac{2}{x} - \dfrac{15}{x^2} = 0$

10. $\dfrac{5x}{x+1} + \dfrac{6}{x+2} = \dfrac{3}{(x+1)(x+2)}$

11. $\dfrac{3}{x} + \dfrac{1}{x-3} = \dfrac{7}{4}$

12. $\dfrac{5}{x} + \dfrac{1}{2x+7} = -\dfrac{2}{3}$

13. $\dfrac{2m}{m-5} + \dfrac{7}{m+1} = 0$

14. $2 + \dfrac{3}{p} = \dfrac{5}{p^2}$

15. $x = \dfrac{1}{x-3} + \dfrac{17}{3}$

16. $\dfrac{2}{x^2} + \dfrac{1}{x} = 1$

Objective 2 Use quadratic equations to solve applied problems.

Solve each problem.

17. The perimeter of a rectangle is 24 inches. The area is 32 square inches. Find the length and the width of the rectangle.

18. Amy rows her boat upstream 6 miles and back in $2\frac{6}{7}$ hours. The speed of the current is 2 miles per hour. How fast can she row?

19. Bill can row 3 miles per hour in still water. It takes him 3 hours and 36 minutes to go 3 miles upstream and return. Find the speed of the current.

20. Two pipes together can fill a large tank in 6 hours. One of the pipes, used alone, takes 5 hours longer than the other to fill the tank. How long would each pipe used alone take to fill the tank?

21. The distance from Appletown to Medina is 45 miles, as is the distance from Medina to Westmont. Karl drove from Westmont to Medina, stopped at Medina for a hamburger, and then drove on to Appletown at 10 miles per hour faster. Driving time for the entire trip was 99 minutes. Find Karl's speed from Westmont to Medina.

22. Two pipes together can fill a large tank in 10 hours. One of the pipes, used alone, takes 15 hours longer than the other to fill the tank. How long would each pipe used alone take to fill the tank?

23. A jet plane traveling at a constant speed goes 1200 miles with the wind, then turns around and travels for 1000 miles against the wind. If the speed of the wind is 50 miles per hour and the total flight takes 4 hours, find the speed of the plane.

24. The sum of the reciprocal of a number and the reciprocal of 5 more than the number is $\frac{11}{24}$. What is the number?

25. A man rode a bicycle for 12 miles and then hiked an additional 8 miles. The total time for the trip was 5 hours. If his rate when he was riding the bicycle was 10 miles per hour faster than his rate walking, what was each rate?

26. The sum of a number and its reciprocal is 5.2. What is the number?

Objective 3   **Solve an equation with radicals by writing it in quadratic form.**

Solve each equation. Check your solutions.

27. $x = \sqrt{x+2}$

28. $\sqrt{7y-10} = y$

29. $\sqrt{2}y = \sqrt{6-y}$

30. $k - \sqrt{8k-15} = 0$

31. $x = \sqrt{\dfrac{x+3}{2}}$

32. $y = \sqrt{\dfrac{1-2y}{8}}$

33. $\sqrt{3}y = \sqrt{28y-49}$

34. $p = \sqrt{\dfrac{7p-1}{12}}$

35. $\sqrt{\dfrac{25r-1}{144}} = r$

36. $y = \sqrt{y+12}$

Objective 4   **Solve an equation that is quadratic in form by substitution.**

Solve each equation. Check your solutions.

37. $x^4 - 25x^2 + 144 = 0$

38. $16m^4 = 25m^2 - 9$

39. $(x+1)^2 = 10(x+1) + 75$

40. $c^4 - 13c^2 + 36 = 0$

41. $(m+5)^2 + 6(m+5) + 8 = 0$

42. $(3-r)^2 = -3(3-r) + 18$

43. $\sqrt{x} = x - 6$

44. $m^{-2} - m^{-1} - 12 = 0$

45. $x^4 = -x^2 + 20$

46. $a^4 + 63 = 16a^2$

47. $t^4 = \dfrac{21t^2 - 5}{4}$

48. $\dfrac{1}{(x+6)^2} - \dfrac{7}{2(x+6)} = -\dfrac{3}{2}$

49. $x^4 - 20x^2 + 36 = 0$

50. $m^4 - 16m^2 = 80$

51. $x^4 - 11x^2 + 24 = 0$

52. $x^4 - 12x^2 + 20 = 0$

53. $m^4 - 10m^2 + 24 = 0$

54. $p^4 - 12p^2 + 27 = 0$

55. $x^4 - 2x^2 + 1 = 0$

56. $x^4 = 3x^2$

57. $\dfrac{1}{9}(r^2 + 3)^2 = r^2 + 1$

58. $36x^4 - 3601x^2 + 100 = 0$

10.4 Mixed Exercises

Solve each problem. Round each answer to the nearest tenth.

59. Andrew can do a job in 3 hours less time than Emily. If they can finish the job together in 10 hours, how long will it take Andrew working alone?

60. Two pipes together can fill a large tank in 15 hours. One of the pipes, used alone, takes 12 hours longer than the other to fill the tank. How long would each pipe used alone take to fill the tank?

61. Cara rowed her boat across and back Lake Bend in 3 hours. If her rate returning was 2 miles per hour less than the rate going, and if the distance each way was 7 miles, find her rate going.

Solve each equation. Check your solutions.

62. $10 - \dfrac{7}{c^2} = -\dfrac{33}{c}$

63. $\dfrac{x-2}{x} + \dfrac{1}{x-1} = \dfrac{5}{6}$

64. $1 + \dfrac{49}{2x} = \dfrac{15}{x+1}$

65. $\dfrac{1}{p} - \dfrac{6}{p^2} = -1$

66. $(m+1)^2 - \dfrac{8}{3}(m+1) = 1$

67. $(t^2 - 3t)^2 = 14(t^2 - 3t) - 40$

68. $(\sqrt{x}+3)^2 = 8(\sqrt{x}+3) - 12$

69. $4m^4 + 12 = 19m^2$

10.5 Formulas and Further Applications

Objective 1   Solve formulas for variables involving squares and square roots.

Solve each equation for the indicated variable. (Leave ± in your answers.)

1. $D = \sqrt{kh}$ for k

2. $F = \dfrac{kl}{\sqrt{d}}$ for d

3. $p = \sqrt{\dfrac{kl}{g}}$ for k

4. $p = \dfrac{yz}{\sqrt{6}}$ for z

5. $s = 30\sqrt{\dfrac{a}{p}}$ for p

6. $a = \sqrt{bc} + 1$ for c

7. $y = \dfrac{1}{2}gt^2$ for t

8. $F = \dfrac{mx}{t^2}$ for t

9. $F = \dfrac{1}{2}kx^2$ for x

Objective 2   Solve applied problems using the Pythagorean formula.

Solve each problem.

10. A 13-foot ladder is leaning against a building. The distance from the bottom of the ladder to the building is 2 feet more than twice the distance from the top of the ladder to the ground. How far is the bottom of the ladder from the building?

11. A ladder is leaning against a building so that the top is 8 feet above the ground. The length of the ladder is 2 feet less than twice the distance of the bottom of the ladder from the building. Find the length of the ladder.

12. Two cars left an intersection at the same time, one heading south, the other heading east. Some time later the car traveling south had gone 18 miles farther than the car headed east. At that time they were 90 miles apart. How far had each car traveled?

13. Two cars left an intersection at the same time, one heading north, the other heading west. Later they were exactly 95 miles apart. The car headed west had gone 38 miles less than twice as far as the car headed north. How far had each car traveled?

14. A child flying a kite has let out 45 feet of string to the kite. The distance from the kite to the ground is 9 feet more than the distance from the child to a point directly below the kite. How high up is the kite?

15. The height of a kite above the ground is 4 feet less than twice the distance from the person flying the kite to a point directly below it. The length of the string to the kite is 68 feet. How high is the kite?

16. The longest side of a right triangle is 4 centimeters longer than the next longest side. The third side is 16 centimeters in length. Find the length of the longest side.

17. The longest side of a right triangle is 2 feet more than the middle side, and the middle side is 1 foot less than twice the shortest side. Find the length of the shortest side.

18. The two shorter sides of a right triangle have lengths that differ by 4 meters. The longest side is 20 meters. Find the shortest side.

19. The longest side of a right triangle is 1 foot more than the middle side. The shortest side is 7 feet. Find the longest side.

20. The width of a rectangle is 14 centimeters. The diagonal is 2 centimeters more than the length. Find the length of the rectangle.

21. The diagonal of a rectangle is 25 inches, and the length is 3 inches more than three times the width. What is the length of the rectangle?

Objective 3   **Solve applied problems using area formulas.**

Solve each problem.

22. A rectangle has a length 1 meter less than twice its width. If 1 meter is cut from the length and added to the width, the figure becomes a square with an area of 16 square meters. Find the dimensions of the original rectangle.

23. The length of a rectangle is 4 inches more than its width. If 2 inches is taken from the length and added to the width, the figure becomes a square with an area of 196 square inches. What are the dimensions of the original figure?

24. The area of a square is 81 square centimeters. If the same amount is added to one dimension and removed from the other, the resulting rectangle has an area 9 centimeters less than the area of the square. How much is added and subtracted?

25. A square has an area of 81 square inches. If the same amount is added to one side and subtracted from an adjacent side, the resulting rectangle has an area of 77 square inches. Find the dimensions of the rectangle.

26. A rectangular piece of cardboard has a length that is 3 inches longer than the width. A square 1.5 inches on a side is cut from each corner. The sides are then turned up to form an open box with a volume of 162 cubic inches. Find the dimensions of the original piece of cardboard.

27. A piece of plastic in the shape of a rectangle has a length 10 inches less than twice the width. A square 4 inches on a side is cut out of each corner and the sides turned up to form an open box with a volume of 160 cubic inches. Find the dimensions of the finished box.

28. An open box is to be made from a rectangular piece of tin by cutting 2-inch squares out of the corners and folding up the sides. The length of the finished box is to be twice the width. The volume of the box will be 100 cubic inches. Find the dimensions of the rectangular piece of tin.

29. To make an open box, 3-centimeter squares are cut from the corners of a rectangular piece of cardboard and the sides folded up. The length of the cardboard is 6 centimeters more than its width. The volume of the finished box will be 120 cubic centimeters. Find the dimensions of the piece of cardboard.

30. The floor of a room 14 feet by 18 feet is to be tiled with a border of even width around the edges. How wide should the border be if the region inside the border is to have an area of 140 square feet?

31. A rectangular garden has an area of 12 feet by 5 feet. A gravel path of equal width is to be built around the garden. How wide can the path be if there is enough gravel for 138 square feet?

32. A picture 9 inches by 12 inches is to be mounted on a piece of mat board so that there is an even width of mat all around the picture. How wide will the matted border be if the area of the mounted picture is 238 square inches?

33. A rug is to fit in a room so that a border of even width is left on all four sides. If the room is 16 feet by 20 feet and the area of the rug is 165 square feet, how wide to the nearest tenth of a foot will the border be?

Objective 4 **Solve applied problems using quadratic functions as models.**

Solve each problem. Round answers to the nearest tenth.

34. A population of microorganisms grows according to the function $p(x) = 100 + .2x + .5x^2$, where x is given in hours. How many hours does it take to reach a population of 250 microorganisms?

35. The charge potential for a certain experiment can be modeled by the function $c(x) = 5 - .1x - .2x^2$, where x is given in minutes. When does the potential equal 4.5?

36. The position of an object moving in a straight line is given by $s(t) = t^2 - 8t$, where s is in feet and t is in seconds. How long will it take the object to move 10 feet?

37. An object is thrown downward from a tower 280 feet high. The distance the object has fallen at time t in seconds is given by $s(t) = 16t^2 + 68t$. How long will it take the object to fall 100 feet?

10.5 Mixed Exercises

Solve each equation for the indicated variable. (Leave ± in your answers.)

38. $pq = t^2 - pt$ for t

39. $xy = m^2 + xm$ for m

40. $p^2q^2 + pkq = k^2$ for q

41. $b^2a^2 + 2bca = c^2$ for a

Solve each problem.

42. A doghouse 2 feet by 4 feet is to be built with a cement path around it of equal width on all sides. The area available for the doghouse and path is 120 square feet. How wide will the path be?

43. The length of a rectangle is 5 centimeters less than the diagonal, and the width is 5 centimeters less than the length. Find the width of the rectangle.

44. The length of a rectangle is 2 inches more than twice the width. The diagonal is 1 inch more than the length. Find the diagonal.

45. A rug is to fit in a room so that a border of even width is left on all four sides. If a room is 12 feet by 15 feet and the area of the rug is 108 square feet, how wide will the border be?

46. The profit from the sale of x items is given by the function $P(x) = 2x^2 - 10x - 100$. What is the minimum number of items that must be sold for the profit to exceed $1000?

10.6 Graphs of Quadratic Functions

Objective 1 **Graph a quadratic function.**

Objective 2 **Graph parabolas with horizontal and vertical shifts.**

Identify the vertex and graph each parabola.

1. $f(x) = x^2 - 2$

2. $f(x) = x^2 + 2$

3. $f(x) = x^2 + 3$

4. $f(x) = x^2 - 4$

5. $f(x) = 2 - x^2$

6. $f(x) = 5 - x^2$

7. $f(x) = (x + 2)^2$

8. $f(x) = (x - 3)^2$

9. $f(x) = (x + 3)^2 - 1$

10. $f(x) = (x - 2)^2 + 1$

Objective 3 **Predict the shape and direction of a parabola from the coefficient of** x^2.

For each quadratic function, tell whether the graph opens up or down and whether the graph is wider, narrower, or the same shape as the graph of $f(x) = x^2$.

11. $f(x) = \dfrac{1}{2}x^2$

12. $f(x) = 3x^2$

13. $f(x) = -2x^2$

14. $f(x) = -\dfrac{4}{3}x^2 - 1$

15. $f(x) = \dfrac{3}{5}x^2 + 5$

16. $f(x) = \dfrac{1}{3}(x - 2)^2$

17. $f(x) = -2(x + 1)^2$

18. $f(x) = -\dfrac{1}{3}(x + 3)^2 - 4$

19. $f(x) = \dfrac{5}{4}(x - 1)^2 + 7$

20. $f(x) = 4 - x^2$

Objective 4 **Find a quadratic function to model data.**

Tell whether a linear or quadratic function would be a more appropriate model for each set of graphed data. If linear, tell whether the slope should be positive or negative. If quadratic, tell whether the coefficient a of x^2 *should be positive or negative.*

21.
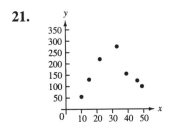

22.
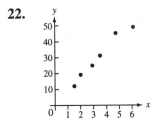

23.
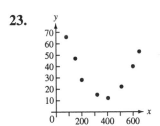

10.6 Mixed Exercises

Identify the vertex and graph each parabola.

24. $f(x) = (x-1)^2$

25. $f(x) = (x+2)^2 + 3$

26. $f(x) = (x-3)^2 - 1$

27. $f(x) = (x+3)^2$

For each quadratic function, tell whether the graph opens up or down and whether the graph is wider, narrower, or the same shape as the graph of $f(x) = x^2$. Find each vertex.

28. $f(x) = 4x^2$

29. $f(x) = -x^2 + 2$

30. $f(x) = \dfrac{1}{3}(x-1)^2$

31. $f(x) = \dfrac{3}{2}(x+1)^2 - 2$

32. $f(x) = -\dfrac{1}{2}(x+3)^2$

33. $f(x) = 4(x-3)^2 + 1$

10.7 More about Parabolas; Applications

Objective 1   Find the vertex of a vertical parabola.

Objective 2   Graph a quadratic function.

Graph each function and find its vertex.

1. $f(x) = x^2 + 6x + 10$

2. $f(x) = x^2 - 6x + 4$

3. $f(x) = -x^2 + 8x - 10$

4. $f(x) = x^2 - 3x + 2$

5. $f(x) = 3x^2 + 6x + 2$

6. $f(x) = -2x^2 + 4x + 1$

7. $f(x) = \frac{1}{2}x^2 + 2x + 3$

8. $f(x) = \frac{5}{4}x^2 + 5x + 3$

Objective 3   Use the discriminant to find the number of *x*-intercepts of a vertical parabola.

Use the discriminant to determine the number of x-intercepts of the graph of each function.

9. $f(x) = 6x^2 - 3x + 1$

10. $f(x) = 3x^2 + 3x + 2$

11. $f(x) = x^2 - 2x + 4$

12. $f(x) = x^2 + 5x + 4$

13. $f(x) = 2x^2 + 4x + 8$

14. $f(x) = 3x^2 + 5x + 1$

15. $f(x) = x^2 - 4x - 4$

16. $f(x) = 5x^2 - 5x + 2$

17. $f(x) = 4x^2 + 12x + 9$

18. $f(x) = x^2 + 5x + 2$

Objective 4   Use quadratic functions to solve problems involving maximum or minimum value.

Solve each problem.

19. A businessman has found that his daily profits are given by $P(x) = -2x^2 + 100x + 2400$, where x is the number of units sold each day. Find the number he should sell daily to maximize his profit. What is the maximum profit?

20. The same businessman has daily costs of $C(x) = x^2 - 50x + 1625$, where x is the number of units sold each day. How many units must be sold to minimize his cost? What is the minimum cost?

21. Jean sells ceramic pots. She has weekly costs of $C(x) = x^2 - 100x + 2700$, where x is the number of pots she sells each week. How many pots should she sell to minimize her costs? What is the minimum cost?

A projectile is fired upward so that its distance (in feet) above the ground t seconds after firing is as given. Find the maximum height it reaches and the number of seconds it takes to reach that height.

22. $s(t) = -16t^2 + 64t$

24. $s(t) = -16t^2 + 48t + 250$

23. $s(t) = -16t^2 + 32t$

25. $s(t) = -16t^2 + 80t + 156$

Solve each problem.

26. The length and width of a rectangle have a sum of 48. What width will produce the maximum area?

27. The length and width of a rectangle have a sum of 64. What width will produce the maximum area?

28. The perimeter of a rectangle is 24. What length will produce the maximum area?

Objective 5   **Graph horizontal parabolas.**

Graph each parabola. Give the domain and range.

29. $x = 2y^2$

33. $x = y^2 + 4y + 4$

30. $x = -2y^2$

34. $x = -y^2 + 4y - 4$

31. $x = -y^2 + 2$

35. $x = y^2 - 4y + 7$

32. $x = y^2 - 3$

36. $3x = y^2 - 6y + 6$

10.7 Mixed Exercises

Graph each parabola and find its vertex.

37. $y = -\dfrac{1}{3}x^2 - 2x - 4$

39. $x = -y^2 - 6y - 10$

38. $x = y^2 + 6y + 5$

40. $y = 2x^2 + 4x - \dfrac{1}{2}$

Use the discriminant to determine the number of x-intercepts of the graph of each function.

41. $f(x) = -x^2 + 5x - 3$

42. $f(x) = 2x^2 - 3x + 2$

Solve each problem.

43. Victor Retzlaff has found that the profits (in dollars) of his video store are approximately given by $p(x) = -x^2 + 16x + 34$, where x is the number of units of videos that he should rent daily to produce the maximum profit. How many units of videos should he rent daily to produce the maximum profit? Find the maximum profit.

44. The perimeter of a rectangle is 16. What length will produce the maximum area?

45. Of all pairs of numbers whose sum is 92, find the pair with the maximum product.

10.8 Quadratic and Rational Inequalities

Objective 1   Solve quadratic inequalities.

Solve each inequality, and graph the solution set.

1. $(x-2)(x+3) \geq 0$

2. $(y-2)(y+3) < 0$

3. $(m-5)(m+2) < 0$

4. $(r+3)(r-2) \geq 0$

5. $k^2 + 7k + 12 > 0$

6. $a^2 - a - 2 \leq 0$

7. $2y^2 < y + 3$

8. $2m^2 - 5m > 12$

9. $6r^2 + 7r + 2 > 0$

10. $8k^2 + 10k > 3$

11. $(x-1)^2 \geq -3$

12. $(x-3)^2 < 0$

13. $(2k+5)^2 \leq -1$

14. $(3p+2)^2 \geq -6$

15. $(2a+9)^2 < 0$

16. $(3x-2)^2 < -1$

Objective 2   Solve polynomial inequalities of degree 3 or more.

Solve each inequality, and graph the solution set.

17. $(x+1)(x-2)(x+4) \leq 0$

18. $(y+2)(y-1)(y-2) < 0$

19. $(k+5)(k-1)(k+3) \leq 0$

20. $(x-1)(x-3)(x+2) \geq 0$

21. $(y-4)(y+3)(y+1) \leq 0$

22. $(p-6)(p-4)(p-2) > 0$

23. $(2x-1)(2x+3)(3x+1) \leq 0$

24. $(4b+1)(6b-1)(3b-7) > 0$

25. $(4q-3)(2q-7)(3q-10) \geq 0$

26. $(z+1)(z-1)(3z-7) < 0$

Objective 3   Solve rational inequalities.

Solve each inequality, and graph the solution set.

27. $\dfrac{7}{x-1} \leq 1$

28. $\dfrac{2}{p+3} \leq 1$

29. $\dfrac{5}{x+1} \geq 1$

30. $\dfrac{4}{m+2} > 3$

31. $\dfrac{-2}{p+1} > 3$

32. $\dfrac{-1}{z-3} \leq 2$

33. $\dfrac{5}{y+2} < 0$

34. $\dfrac{3}{a+4} > 0$

35. $\dfrac{6}{3r-2} \geq 1$

36. $\dfrac{4}{3q+5} \geq -3$

37. $\dfrac{-5}{2x-3} \leq 2$

38. $\dfrac{-3}{4m-3} \geq 1$

39. $\dfrac{y}{y+1} \geq 3$

40. $\dfrac{r}{r-2} \geq 4$

41. $\dfrac{2p-1}{3p+1} \leq 1$

42. $\dfrac{z+2}{z-3} \leq 2$

43. $\dfrac{5}{x-3} \leq -1$

44. $\dfrac{4z}{3z-5} < -3$

10.8 Mixed Exercises

Solve each inequality, and graph the solution set.

45. $15k^2 + 2 \le 11k$

46. $6z^2 \ge 11z + 10$

47. $4x^2 - 9 \le 0$

48. $16r^2 - 9 \ge 0$

49. $8p^2 + 2p > 1$

50. $6m^2 + 17m < 3$

51. $(m - 2)(m + 1)(m + 3) \ge 0$

52. $(y + 2)(y - 2)(y - 3) < 0$

53. $(2r - 1)(r - 3)(r + 1) < 0$

54. $(4a - 3)(a - 1)(a - 3) \ge 0$

55. $(3k - 1)(2k + 3)(k - 2) > 0$

56. $(5m - 4)(2m + 5)(m + 1) \le 0$

57. $\dfrac{m - 3}{m} \ge 3$

58. $\dfrac{r + 2}{r} \le 5$

59. $\dfrac{2p - 3}{3p} \le -1$

60. $\dfrac{4k - 3}{2k} \ge -2$

61. $(y + 7)^2 + 6 \le 0$

62. $(2x - 1)^2 + 3 \ge 0$

Chapter 11

EXPONENTIAL AND LOGARITHMIC FUNCTIONS

11.1 Inverse Functions

Objective 1 Decide whether a function is one-to-one and, if it is, find its inverse.

If the function is one-to-one, find its inverse.

1. $\{(1, 0), (2, 0), (3, 5), (4, 1)\}$

2. $\{(-1, 1), (1, -1), (2, 1)\}$

3. $\{(2, -1), (-2, 1), (1, 3), (-1, -3)\}$

4. $\{(6, -3), (4, -2), (2, -1), (0, 0)\}$

5. $\{(4, 0), (2, 3), (0, 0), (3, 5)\}$

6. $\{(1, 5), (2, 5), (3, 5)\}$

7. $\{(-1, 1), (-2, 2), (-3, 3)\}$

8. $\{(-3, 1), (-2, 2), (-1, 3), (0, 4)\}$

9. $\{(3, 2), (-3, -2), (2, 3), (-2, -3)\}$

10. $\{(5, -1), (4, 7), (6, 3), (3, 3)\}$

Objective 2 Use the horizontal line test to determine whether a function is one-to-one.

Use the horizontal line test to determine whether each function is one-to-one.

11.

14.
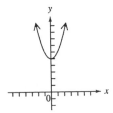

17.
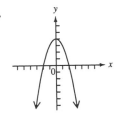

12.
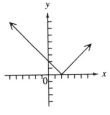

15.
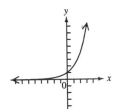

18.
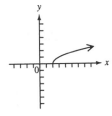

13.
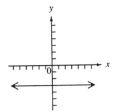

16.
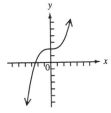

Objective 3 Find the equation of the inverse of a function.

If the function is one-to-one, find its inverse.

19. $f(x) = 2x - 5$

20. $f(x) = 3x - 5$

21. $f(x) = x^2 - 1$

22. $f(x) = 1 - 2x^2$

23. $f(x) = \sqrt{x - 1}, \ x \geq 1$

24. $f(x) = 2\sqrt{3x}, \ x \geq 0$

25. $f(x) = x^3 - 1$

26. $f(x) = 2x^3 - 3$

28. $f(x) = \dfrac{3}{x-1}$

27. $f(x) = \dfrac{x^2 + 3}{2}$

Objective 4 **Graph f^{-1} from the graph of f.**

If the function is one-to-one, graph the function f and its inverse f^{-1} on the same set of axes.

29.

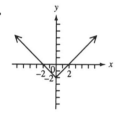

32.

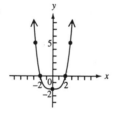

35.

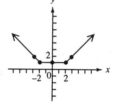

30.

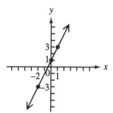

33.

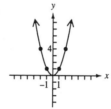

36.

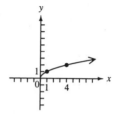

31.

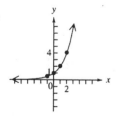

34.

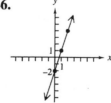

11.1 Mixed Exercises

If the function is one-to-one, find its inverse.

37. $\{(3, 5), (2, 9), (4, 7)\}$

38. $\{(0, 0), (1, 1), (-1, -1), (2, 2), (-2, -2)\}$

39. $\{(2, 4), (-1, 1), (0, 0), (1, 1), (2, 4)\}$

40. $\{(-3, -1), (-2, 0), (-1, 1), (0, 2)\}$

41. $f(x) = 4 - 2x$

42. $f(x) = \sqrt{x + 2},\ x \geq -2$

43. $f(x) = 2x^2 + 3$

44. $f(x) = x^3 - 5$

Use the horizontal line test to determine whether each function is one-to-one. If it is, graph the function f and its inverse f^{-1} on the same set of axes.

45.

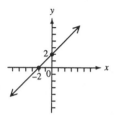

46.

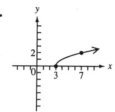

11.2 Exponential Functions

Objective 1 Define exponential functions.

Decide whether or not each function defines an exponential function.

1. $f(x) = 2^x$

2. $f(x) = x^2$

3. $f(x) = x + 2$

4. $f(x) = 3^x$

5. $f(x) = (-3)^x$

6. $f(x) = (x+1)^3$

7. $f(x) = 2^{x+1}$

8. $f(x) = 3^{2x}$

9. $f(x) = 2x^3$

10. $f(x) = 1^x$

Objective 2 Graph exponential functions.

Graph each exponential function.

11. $f(x) = 3^x$

12. $f(x) = -4^x$

13. $f(x) = 2^{-x}$

14. $f(x) = \left(\dfrac{1}{8}\right)^x$

15. $f(x) = 2^{1-x}$

16. $f(x) = -2^{x-2}$

Objective 3 Solve exponential equations of the form $a^x = a^k$ for x.

Solve each equation.

17. $16^x = 64$

18. $27^k = 9$

19. $25^p = 625$

20. $4^{2x} = 8$

21. $9^y = 3$

22. $10^{2x} = 100$

23. $4^{k+2} = 32$

24. $25^{2x-1} = 5$

25. $25^{1-t} = 5$

26. $100^{2+t} = 1000$

Objective 4 Use exponential functions in applications involving growth or decay.

Solve each problem.

27. The population of Canadian geese that spend the summer at Gemini Lake each year has been growing according to the function

$$f(x) = 56(2)^{.2x},$$

where x is the time in years from 1978. Find the number of geese in 1994.

28. The diameter in inches of a tree during a certain period grew according to the function

$$f(x) = 2.5(9)^{.05x},$$

where x was the number of years after the start of this growth period. Find the diameter of the tree after 10 years.

29. A culture of a certain kind of bacteria grows according to the function

$$f(x) = 3650(2)^{.8x},$$

where x is the time in hours after 12 noon. Find the number of bacteria in the culture at 12 noon.

30. An industrial city in Pennsylvania has found that its population is declining according to the function

$$f(x) = 70,000(2)^{-.01x},$$

where x is the time in years from 1900. What is the city's anticipated population in the year 2000?

31. A sample of a radioactive substance with mass in grams decays according to the function

$$f(x) = 100(10)^{-.2x},$$

where x is the time in hours after the original measurement. Find the mass of the substance after 10 hours.

32. When a bactericide is placed in a certain culture of bacteria, the number of bacteria decreases according to the function

$$f(x) = 3200(4)^{-.1x},$$

where x is the time in hours. Find the number of bacteria in the culture after 20 hours.

33. Suppose the number of bacteria present in a certain culture after t minutes is given by the function

$$Q(t) = 500(2)^{.5t}.$$

Find the number of bacteria present after 2 minutes.

34. The population of Evergreen Park is now 16,000. The population t years from now is given by the function

$$P(t) = 16,000(2)^{t/10}.$$

What will the population be 40 years from now?

11.2 Mixed Exercises

Solve each equation.

35. $4^x = 8$

36. $25^{-2x} = 3125$

37. $16^{-x+1} = 8$

38. $5^{-x} = \dfrac{1}{5}$

39. $\left(\dfrac{1}{3}\right)^x = 27$

40. $\left(\dfrac{3}{4}\right)^x = \dfrac{16}{9}$

Graph each exponential function..

41. $f(x) = 4^{2x-3}$

42. $f(x) = -3^{-x}$

Solve each problem.

43. Corinna's savings grows according to the function

$$A(t) = P(1.01)^{4t},$$

where P is the amount of her original deposit and t is the number of years since the deposit was made. If $P = \$10,000$, how much will she have in 25 years?

44. The production of an oil well, in barrels, is decreasing according to the function

$$f(t) = 1,000,000(2)^{-.4t},$$

where t is the number of years after the well was drilled. Find the production after 5 years.

11.3 Logarithmic Functions

Objective 1 **Define a logarithm.**

Simplify. (Example: $\log_3 9 = 2.$)

1. $\log_2 8$

2. $\log_8 64$

3. $\log_3 \dfrac{1}{3}$

4. $\log_3 \sqrt{3}$

5. $\log_{1/2} 4$

6. $\log_{10} .0001$

7. $\log_7 \sqrt{7}$

8. $\log_4 2$

9. $\log_5 \dfrac{1}{25}$

10. $\log_{81} 27$

Objective 2 **Convert between exponential and logarithmic forms.**

Complete the chart.

	Exponential Form	Logarithmic Form
11.	$3^2 = 9$	
12.	$5^{1/3} = \sqrt[3]{5}$	
13.		$\log_4 \dfrac{1}{16} = -2$
14.		$\log_{16} 2 = \dfrac{1}{4}$
15.	$10^{-2} = \dfrac{1}{100}$	
16.		$\log_5 25 = 2$
17.		$\log_9 3 = \dfrac{1}{2}$
18.	$9^{1/2} = 3$	
19.		$\log_{10} .001 = -3$
20.	$2^{-7} = \dfrac{1}{128}$	

Objective 3   **Solve logarithmic equations of the form** $\log_a b = k$ **for** $a, b,$ **or** k.

Solve each equation.

21. $\log_2 64 = p$

22. $\log_5 x = -1$

23. $\log_m 25 = 2$

24. $\log_k 27 = 3$

25. $\log_2 t = -5$

26. $\log_4 16 = y$

27. $\log_n .01 = -2$

28. $\log_{1/2} r = -2$

29. $\log_4 x = 0$

30. $\log_{1/4} \dfrac{1}{4} = p$

Objective 4   **Define and graph logarithmic functions.**

Graph each logarithmic function.

31. $y = \log_2 x$

32. $y = \log_9 x$

33. $y = -\log_4 x$

34. $y = \log_{1/2} x$

35. $y = \log_{1/9} x$

36. $y = -\log_{1/4} x$

37. $y = \log_{10} x$

38. $y = \log_2 (-x)$

Objective 5   **Use logarithmic functions in applications of growth or decay.**

Solve each problem.

39. After black squirrels were introduced to Williams Park, their population grew according to the function

$$f(x) = 10\log_5 (x+20),$$

where x is the number of months after the squirrels were introduced. Find the number of squirrels after 5 months.

40. A manufacturer receives revenue in dollars for selling x units of an item according to the function
$$f(x) = 200\log_3 (x+1).$$

Find the revenue for selling 26 units.

41. Under certain conditions, the velocity v of the wind in centimeters per second is given by

$$v = 300\log_2 \left(\frac{10x}{7}\right),$$

where x is the height in centimeters above the ground. Find the wind velocity at 11.2 centimeters above the ground.

42. A decibel is a measure of the loudness of a sound. A very faint sound is assigned an intensity of I_0, then another sound is given an intensity I found in terms of I_0, the faint sound. The decibel rating of the sound is given in decibels by

$$d = 10\log_{10} \frac{I}{I_0}.$$

Find the decibel rating of rock music that has intensity $I = 100,000,000,000 I_0$.

43. The number of students not completing intermediate algebra is given by the function

$$f(s) = 4\log_7(8s + 9),$$

where s is the number of sections of the class that is offered. If there are 5 sections of intermediate algebra this semester, how many students will not complete the course?

44. The number of fish in an aquarium is given by the function

$$f(t) = 10\log_2(3t + 2),$$

where t is time in months. Find the difference in the number of fish present between $t = 0$ and $t = 10$.

45. After black squirrels were introduced to Sherman Park, their population grew according to the function

$$f(x) = 12\log_5(2x + 5),$$

where x is the number of months after the squirrels were introduced. How many more squirrels were there in the park 10 months after being introduced than there were originally?

46. A company analyst has found that the number of applicants for new mortgages after a major advertising blitz is given by the function

$$A(x) = 50\log_2(2x + 2),$$

where x is time in weeks after the blitz was started. Find the number of applicants for $x = 7$.

47. Sales (in thousands) of a new product are approximated by

$$S = 125 + 20\log_2(30t + 4) + 30\log_4(35t - 6),$$

where t is the number of years after the product is introduced. Find the total sales 2 years after the product is introduced.

48. A population of mites in a laboratory is growing according to the function

$$p = 50\log_3(20t + 7) - 25\log_9(80t + 1),$$

where t is the number of days after a study is begun. Find the number of mites present 1 day after the beginning of the study.

11.3 Mixed Exercises

49. Write $\left(\frac{1}{2}\right)^{-3} = 8$ in logarithmic form.

50. Write $\log_5 .0016 = -4$ in exponential form.

Solve each equation.

51. $x = \log_{32} 8$

53. $\log_5 1 = x$

55. Graph the function
$y = \log_3 3x.$

52. $\log_{1/3} r = -4$

54. $\log_a 4 = \dfrac{1}{2}$

Solve each problem.

56. The population of foxes in an area t months after the foxes were introduced there is approximated by the function

$$F(t) = 500 \log_{10}(2t + 3).$$

Find the number of foxes in the area when the foxes were first introduced into the area.

57. The number of fish in an aquarium is given by the function

$$f(t) = 8 \log_5(2t + 5),$$

where t is time in months. Find the number of fish present when $t = 10$.

58. A company analyst has found that total sales in thousands of dollars after a major advertising campaign are given by

$$S(x) = 100 \log_2(x + 2),$$

where x is time in weeks after the campaign was introduced. Find the sales when the campaign was introduced.

11.4 Properties of Logarithms

Objective 1   **Use the product rule for logarithms.**

Use the product rule to express each logarithm as a sum of logarithms, or as a single number if possible.

1. $\log_3 (6)(5)$

2. $\log_2 (5)(3)$

3. $\log_7 5m$

4. $\log_2 6xy$

5. $\log_6 6r$

6. $\log_3 2p$

Use the product rule to express each sum as a single logarithm.

7. $\log_4 7 + \log_4 3$

8. $\log 4 + \log 3$

9. $\log_7 11y + \log_7 2y + \log_7 3y$

10. $\log_7 8r^2 + \log_7 5r^2 + \log_7 3r$

Objective 2   **Use the quotient rule for logarithms.**

Use the quotient rule for logarithms to express each logarithm as a difference of logarithms, or as a single number if possible.

11. $\log_2 \dfrac{7}{9}$

12. $\log_4 \dfrac{5}{8}$

13. $\log_3 \dfrac{m}{n}$

14. $\log \dfrac{p}{r}$

15. $\log_6 \dfrac{k}{3}$

16. $\log_3 \dfrac{10}{x}$

17. $\log_2 \dfrac{8}{m}$

18. $\log_5 \dfrac{5}{x}$

Use the quotient rule for logarithms to express each difference as a single logarithm.

19. $\log_2 7q^4 - \log_2 5q^2$

20. $\log 9x^3 - \log 3x^2$

21. $\log_7 60r^3 - \log_7 100r^7$

22. $\log_9 40y^5 - \log_9 20y^7$

Objective 3   **Use the power rule for logarithms.**

Use the power rule for logarithms to rewrite each logarithm.

23. $\log_5 3^2$

24. $\log_3 4^3$

25. $\log_2 5^3$

26. $\log_m 2^7$

27. $\log_b \sqrt{5}$

28. $\log_3 \sqrt[3]{7}$

29. $\log_2 \sqrt{2}$

30. $\log_4 \sqrt[3]{4}$

31. $\log_2 \sqrt[3]{8}$

32. $\log_5 125^{1/3}$

Objective 4   **Use properties to write alternative forms of logarithmic expressions.**

Use the properties of logarithms to express each logarithm as a sum or difference of logarithms, or as a single number if possible.

33. $\log_2 4p^3$

34. $\log_3 9x^3$

35. $\log_a \sqrt[3]{2k}$

36. $\log_b \dfrac{2r}{r-1}$

37. $\log_2 \dfrac{4}{3}$

38. $\log_3 \dfrac{5}{9}$

39. $\log_5 \dfrac{7m^3}{8y}$

40. $\log_7 \dfrac{8r^7}{3a^3}$

Use the properties of logarithms to express each sum or difference of logarithms as a single logarithm, or as a single number if possible.

41. $\log 2x + \log 7x$

42. $\log_a 2r + \log_a 4r^2$

43. $\log_b 3pq - \log_b 2p^2$

44. $\log 4k^2 j - \log 3kj^2$

45. $\log_4 10y + \log_4 3y - \log_4 6y^3$

46. $\log_6 14m + \log_6 7m^2 - \log_6 14m^4$

47. $\log_2 (x-1) + \log_2 (x+1) - \log_2 (x^2 - 1)$

48. $\log_6 (r+4) + \log_6 (r-3) - \log_6 (r^2 + r - 12)$

11.4 Mixed Exercises

Use the properties of logarithms to express each logarithm as a sum or difference of logarithms, or as a single number if possible.

49. $\log_2 8p$

50. $\log_3 \sqrt{27}$

51. $\log_4 \dfrac{4}{9}$

52. $\log_5 k^4$

53. $\log_2 32^{2/5}$

54. $\log_5 \sqrt{3p}$

55. $\log_7 \dfrac{3}{7}$

56. $\log_4 \dfrac{3m}{m+2}$

Use the properties of logarithms to express each sum of difference of logarithms as a single logarithm, or as a single number if possible.

57. $\log_2 6y^3 - \log_2 3y^3$

58. $\log_3 2q + \log_3 5q^3$

59. $\log_5 8y + \log_5 2y$

60. $\log_{10} 1000p^5 - \log_{10} 100p^5$

61. $2\log_5 m + 3\log_5 m^2 - 4\log_5 m^3$

62. $2\log_2 y^2 + \log_2 y - 2\log_2 y^3$

11.5 Common and Natural Logarithms

Objective 1 Evaluate common logarithms using a calculator.

Use a calculator to find each logarithm. Give an approximation to four decimal places.

1. log 57.23
2. log 8
3. log 843.71
4. log .091419
5. log 280,037
6. log 798.886

7. log .00003184
8. log 61.000958
9. log .000958
10. log 87,123
11. log 22
12. log .3501

13. log 767
14. log 5489.62
15. log .000829
16. log .001
17. log 1,031,057
18. log 4.0014

Objective 2 Use common logarithms in applications.

Find the pH of solutions with the given hydronium ion concentrations. Round answers to the nearest tenth.

19. 2.8×10^{-6}
20. 5.6×10^{-8}

21. 2.1×10^{-7}
22. 1.7×10^{-9}

23. 6.2×10^{-5}
24. 7.4×10^{-11}

Find the hydronium ion concentration of solutions with the given pH values.

25. 2.9
26. 3.4

27. 5.2
28. 1.3

29. 6.5
30. 10.2

Objective 3 Evaluate natural logarithms using a calculator.

Find each natural logarithm. Give an approximation to four decimal places.

31. ln .12
32. ln 100
33. ln 6
34. ln 428

35. ln .013
36. ln 69
37. ln 4
38. ln .102

39. ln 874
40. ln 76.3
41. ln .01
42. ln .00214

Objective 4 Use natural logarithms in applications.

Solve each problem.

The population of a small town from 1975–1995 is approximated by the function

$$P(t) = 600e^{.01t},$$

where $t = 0$ represents 1975. Find the population in the given years.

43. 1975
44. 1978
45. 1985
46. 1995

A radioactive substance, in grams, is decaying so that the amount present at time t in days is given by the function

$$Q(t) = 100e^{-.03t}.$$

Find the amount present to the nearest tenth of a gram after the given number of days.

47. Initially   **48.** 10   **49.** 30   **50.** 60

The number of bacteria in a certain culture is approximated by

$$B(t) = 10,000e^{.05t},$$

where t is the time in hours. Find the population present to the nearest hundred after the given number of hours.

51. Initially   **52.** 3   **53.** 4   **54.** 24

11.5 Mixed Exercises

Use a calculator to find each logarithm. Give an approximation to four decimal places.

55. log .093621   **56.** ln 50   **57.** ln .000806   **58.** log 60,183.006

Solve each problem.

59. Find the pH of a substance with hydronium ion concentration of 3.9×10^{-9}. Round the answer to the nearest tenth.

60. Find the hydronium ion contraction of a substance with pH of 4.8. Round the answer to the nearest tenth.

61. Suppose a certain collection of termites is growing according to the function

$$f(t) = 3000e^{.04t},$$

where t is measured in months. If there are 3000 present on January 1, how many will be present on July 1?

62. Suppose the population of a small town is approximated by the function

$$p(t) = 10,000e^{.05t},$$

where t represents the time in years. The population at time $t = 0$ was 10,000. Find the population to the nearest thousand at time $t = 14$.

11.6 Exponential and Logarithmic Equations; Further Applications

Objective 1   Solve equations involving variables in the exponents.

Solve each equation. Give solutions to three decimal places.

1. $27^x = 5$

2. $32^y = 6$

3. $7^m = 11$

4. $12^{-p} = 32$

5. $4^{m-3} = 6$

6. $8^{4-x} = 3$

7. $2^{3y-9} = 7$

8. $4^{p+3} = 10$

9. $6^{2-r} = 50$

10. $5^{2k-1} = 17$

Objective 2   Solve equations involving logarithms.

Solve each equation. Give the exact solution.

11. $\log(p-2) = \log 3$

12. $\log(2k+1) = \log 7$

13. $\log_2(x+1) - \log_2 x = \log_2 5$

14. $\log_3(x-1) + \log_3 x = \log_3 6$

15. $\log(-y) + \log 4 = \log(2y+5)$

16. $\log_m 8 = 3$

17. $\log_p 10 = 4$

18. $\log_a 25 = \dfrac{1}{2}$

19. $\log_3 a = \log_3(a-1) + 2$

20. $\log_4 u = 1 - \log_4(u+3)$

Objective 3   Solve applications of compound interest.

Solve each problem.

Find the final amount owed for each of the following borrowed amounts if interest is compounded annually. Use

$$A = P\left(1 + \frac{r}{n}\right)^{nt},$$

where A is the amount owed, P is the amount borrowed, r is the interest rate, $n = 1$, and t is the time in years.

21. $1000 for 3 years at 8%

22. $25,000 for 5 years at 10%

23. $5600 for 8 years at 11%

24. $2700 for 10 years at 9%

25. $3950 for 5 years at 7%

26. $47,200 for 9 years at 10%

Objective 4 Solve applications involving base e exponential growth and decay.

Solve each problem.

27. Radioactive strontium decays according to the function

$$y = y_0 e^{-.0239t},$$

where t is the time in years. If an initial sample contains $y_0 = 15$ g of radioactive strontium, how many grams will be present after 25 years? Round to the nearest hundredth of a gram.

28. What is the half-life of radioactive strontium in Exercise 27? Round to the nearest year.

29. A sample of 500 g of lead-210 decays to polonium-210 according to the function

$$A(t) = 500e^{-.032t},$$

where t is the time in years. How much lead will be left in the sample after 20 years? Round to the nearest gram.

30. What is the half-life of the initial sample of lead in Exercise 29? Round to the nearest year.

Objective 5 Use the change-of-base rule.

Use the change-of-base rule to find each logarithm. Give approximations to four decimal places.

31. $\log_6 3$

32. $\log_2 10$

33. $\log_7 28$

34. $\log_5 180$

35. $\log_3 142$

36. $\log_{16} 27$

37. $\log_2 14$

38. $\log_5 243$

39. $\log_{1/2} 6$

40. $\log_{1/4} 11$

41. $\log_{1/7} 12$

42. $\log_{2/3} 5$

43. $\log_{1/2} \dfrac{3}{4}$

44. $\log_{1/3} \dfrac{1}{6}$

11.6 Mixed Exercises

Solve each equation. Give solutions to three decimal places.

45. $5^x = 16$

46. $3^{-q} = 2$

47. $10^{k-2} = 24$

48. $11^{5-w} = 7$

49. $\log_t 100 = \dfrac{2}{3}$

50. $\log_2 x + \log_2 (3x-1) = 1$

Find the final amount owed for each of the following borrowed amounts if interest is compounded annually.

51. $72,600 for 4 years at 12%

52. $32,800 for 7 years at 11%

Suppose that over several years a certain average annual rate of inflation is compounded annually. The formula for the price of an item can be found using the formula

$$A = P(1+r)^t,$$

where r is the rate of inflation and A is the price in dollars of the item after t years, if it cost P dollars when t = 0. Find the number of years, to the nearest year, that it takes an item to double in price for the following rates of inflation.

53. 4%

54. 11%

Use the change-of-base rule to find each logarithm. Give approximations to four decimal places.

55. $\log_8 12$

56. $\log_4 8$

Chapter 12

NONLINEAR FUNCTIONS, CONIC SECTIONS, AND NONLINEAR SYSTEMS

12.1   Additional Graphs of Functions; Composition

Objective 1   Recognize the graphs of the elementary functions defined by $|x|, \frac{1}{x}$, and $\sqrt{x}$, and graph their translations.

Graph each function.

1. $f(x) = |x-2| + 3$

2. $f(x) = \sqrt{x+3}$

3. $f(x) = \dfrac{1}{x-1}$

4. $f(x) = -|x+3| - 2$

5. $f(x) = \sqrt{5-x}$

6. $f(x) = \dfrac{1}{x} + 3$

7. $f(x) = |x-3| - 2$

8. $f(x) = \sqrt{x} + 3$

Objective 2   Recognize and graph step functions

Objective 3   Perform operations on functions

Let $f(x) = 2x^2$ *and* $g(x) = x^2 + 1$. *Perform the following operations.*

9. $(f+g)(x)$

10. $(f-g)(x)$

11. $(fg)(x)$

12. $\left(\dfrac{f}{g}\right)(x)$

Objective 4   Find the composition of functions.

Let $f(x) = x^2 + 3$, $g(x) = 3x+2$, *and* $h(x) = x+4$. *Find each composite function.*

13. $(h \circ g)(2)$

14. $(f \circ g)(1)$

15. $(g \circ f)(3)$

16. $(h \circ f)(4)$

17. $(f \circ h)(-3)$

18. $(f \circ g)(x)$

19. $(g \circ h)(x)$

20. $(f \circ h)(x)$

21. $(g \circ f)(x)$

22. $(h \circ g)(x)$

12.1 Mixed Exercises

Graph each function.

23. $f(x) = |x-4| + 2$

24. $f(x) = -\sqrt{x-3} - 3$

25. $f(x) = \dfrac{1}{x-3}$

26. $f(x) = -|x-3|+3$ **27.** $f(x) = \sqrt{4-x^2}$ **28.** $f(x) = -\sqrt{1-x^2}$

Let $f(x) = x-3$ and $g(x) = x^2 + 6$. **Find each composite function or perform the indicated operation.**

29. $(f \circ g)(2)$

30. $(g \circ f)(5)$

31. $(f \circ g)(x)$

32. $(g \circ f)(x)$

33. $(f - g)(-1)$

34. $\left(\dfrac{g}{f}\right)(2)$

12.2   The Circle and the Ellipse

Objective 1 Find the equation of a circle given the center and radius.

Find the equation of a circle with the given conditions.

1. center: $(-3, 2)$; radius: 5

2. center: $(1, 4)$; radius: 2

3. center: $(0, 5)$; radius: 3

4. center: $(6, 2)$; radius: 3

5. center: $(-5, 4)$; radius: 4

6. center: $(7, 1)$; radius: 2

7. center: $(3, -4)$; radius: 5

8. center: $(2, 2)$; radius: 6

9. center: $(1, 3)$; radius: 5

10. center: $(-2, -2)$; radius: 3

Objective 3 Determine the center and radius of a circle given its equation.

Find the center and radius of each circle. In Exercises 11 and 12, sketch each graph.

11. $x^2 + y^2 - 4x + 8y + 11 = 0$

12. $x^2 + y^2 + 6x - 4y + 12 = 0$

13. $x^2 + y^2 - 6x + 10y = 30$

14. $x^2 + y^2 - 4x - 2y = 31$

15. $x^2 + y^2 + 4x + 6y - 3 = 0$

16. $x^2 + y^2 - 10x + 12y + 52 = 0$

17. $x^2 + y^2 - 8x - 2y + 15 = 0$

18. $x^2 + y^2 - 4x + 8y + 11 = 0$

19. $2x^2 + 2y^2 + 4y - 8x = 4$

20. $3x^2 + 3y^2 + 12y + 30x = 21$

Objective 3 Recognize the equation of an ellipse.

Objective 4 Graph ellipses.

Graph each ellipse.

21. $\dfrac{x^2}{9} + \dfrac{y^2}{49} = 1$

22. $\dfrac{x^2}{25} + \dfrac{y^2}{4} = 1$

23. $\dfrac{x^2}{25} + \dfrac{y^2}{36} = 1$

24. $\dfrac{x^2}{4} + \dfrac{y^2}{9} = 1$

25. $\dfrac{x^2}{16} + \dfrac{y^2}{25} = 1$

26. $\dfrac{x^2}{36} + \dfrac{y^2}{9} = 1$

27. $\dfrac{x^2}{25} + \dfrac{y^2}{64} = 1$

28. $\dfrac{x^2}{4} + \dfrac{y^2}{16} = 1$

12.2 Mixed Exercises

Find the equation of a circle with the given conditions.

29. center: $(-2, -4)$; radius: 5

30. center: $(0, 3)$; radius: $\sqrt{2}$

Find the center and radius of each circle.

31. $x^2 + y^2 + 8x + 4y - 29 = 0$

32. $4x^2 + 4y^2 - 24x + 16y + 43 = 0$

Graph each ellipse.

33. $\dfrac{x^2}{16} + \dfrac{y^2}{49} = 1$

34. $\dfrac{x^2}{25} + \dfrac{y^2}{81} = 1$

12.3   The Hyperbola and Other Functions Defined by Radicals

Objective 1 Recognize the equation of a hyperbola.

Objective 2 Graph hyperbolas by using the asymptotes.

Graph each hyperbola.

1. $\dfrac{x^2}{9} - \dfrac{y^2}{16} = 1$ 　　　**3.** $\dfrac{y^2}{4} - \dfrac{x^2}{9} = 1$ 　　　**5.** $\dfrac{x^2}{36} - \dfrac{y^2}{49} = 1$ 　　　**7.** $\dfrac{x^2}{25} - \dfrac{y^2}{4} = 1$

2. $\dfrac{x^2}{25} - \dfrac{y^2}{9} = 1$ 　　　**4.** $\dfrac{y^2}{25} - \dfrac{x^2}{16} = 1$ 　　　**6.** $\dfrac{y^2}{4} - \dfrac{x^2}{4} = 1$ 　　　**8.** $\dfrac{x^2}{25} - \dfrac{y^2}{81} = 1$

Objective 3 Identify conic sections by their equations.

Identify each of the following as the equation of a parabola, a circle, an ellipse, or a hyperbola.

9. $x^2 = y^2 + 9$ 　　　　　　**13.** $25y^2 + 100 = 4x^2$ 　　　　　**17.** $5x^2 = 25 - 5y^2$

10. $2x^2 + 3y^2 = 6$ 　　　　　**14.** $16x^2 + 9y = 144$ 　　　　　**18.** $y^2 = 36 - 36x^2$

11. $2x + y^2 = 16$ 　　　　　　**15.** $16x^2 + 16y^2 = 64$

12. $4x^2 - 9y^2 = 36$ 　　　　　**16.** $2x^2 + 2y^2 = 8$

Objective 4 Graph certain square root functions.

Sketch each graph.

19. $f(x) = \sqrt{36 - x^2}$ 　　　　　　　**23.** $f(x) = \sqrt{1 + \dfrac{x^2}{4}}$ 　　　　　　**26.** $f(x) = -5\sqrt{1 - \dfrac{x^2}{9}}$

20. $f(x) = \sqrt{25 - x^2}$

21. $f(x) = -\sqrt{4 - x^2}$ 　　　　　　**24.** $f(x) = -3\sqrt{1 + \dfrac{x^2}{25}}$

22. $f(x) = -\sqrt{9 - x^2}$ 　　　　　　**25.** $f(x) = \sqrt{9 - 9x^2}$

12.3 Mixed Exercises

Identify each of the following as the equation of a parabola, a circle, an ellipse, or a hyperbola.

27. $3x^2 - 3y = 9$ 　　　　　　**29.** $x^2 = 49 - y^2$ 　　　　　　**31.** $3x^2 = 3y^2 + 1$

28. $3x^2 + 3y^2 = 1$ 　　　　　**30.** $x^2 = 16 - y$ 　　　　　　**32.** $x + y^2 = 16$

12.4 Nonlinear Systems of Equations

Objective 1   Solve a nonlinear system by substitution.

Solve each system by the substitution method.

1. $x^2 + y^2 = 17$
$\quad 2x = y + 9$

2. $2x^2 - y^2 = -1$
$\quad 2x + y = 7$

3. $4x^2 + 3y^2 = 7$
$\quad 2x - 5y = -7$

4. $x^2 = 2y^2 + 2$
$\quad y = 3x + 7$

5. $y = x^2 - 3x - 8$
$\quad x = y + 3$

6. $x = y^2 + 5y$
$\quad 3y = x$

7. $xy = -6$
$\quad x + y = 1$

8. $xy = 24$
$\quad y = 2x + 2$

9. $xy = -10$
$\quad 2x - y = 9$

10. $xy = 10$
$\quad x + y = 7$

Objective 2   Use the elimination method to solve a system with two second-degree equations.

Solve each system by the the elimination method.

11. $x^2 + y^2 = 10$
$\quad 2x^2 - y^2 = -7$

12. $2x^2 + y^2 = 54$
$\quad x^2 - 3y^2 = 13$

13. $2x^2 - 3y^2 = -19$
$\quad 4x^2 + y^2 = 25$

14. $x^2 - y^2 = 3$
$\quad 2x^2 + y^2 = 9$

15. $x^2 + 2y^2 = 11$
$\quad 2x^2 - y^2 = 17$

16. $3x^2 + 2y^2 = 30$
$\quad 2x^2 + y^2 = 17$

17. $5x^2 + y^2 = 6$
$\quad 2x^2 - 3y^2 = -1$

18. $4x^2 - 3y^2 = -8$
$\quad 2x^2 + y^2 = 5$

19. $3x^2 - 3y^2 = 9$
$\quad 4x^2 + y^2 = 17$

20. $3x^2 - 2y^2 = 12$
$\quad x^2 + 3y^2 = 4$

Objective 3   Solve a system that requires a combination of methods.

Solve each system.

21. $x^2 + xy + y^2 = 43$
$\quad x^2 + 2xy + y^2 = 49$

22. $x^2 + xy - y^2 = 5$
$\quad -x^2 + xy + y^2 = -1$

23. $x^2 + 2xy + 3y^2 = 6$
$\quad x^2 + 4xy + 3y^2 = 8$

24. $2x^2 + 3xy - 2y^2 = 50$
$\quad x^2 - 4xy - y^2 = -41$

25. $4x^2 - 2xy + 4y^2 = 64$

    $x^2 \qquad + \; y^2 = 13$

26. $5x^2 - xy + 5y^2 = 89$

    $x^2 + \qquad y^2 = 17$

27. $3x^2 - 4xy + 2y^2 = \; 59$

    $-3x^2 + 5xy - 2y^2 = -65$

28. $x^2 + 3xy + 2y^2 = 12$

    $-x^2 + 8xy - 2y^2 = 10$

29. $x^2 + 5xy - y^2 = 20$

    $x^2 - 2xy - y^2 = -8$

30. $2x^2 + \; xy + y^2 = \; 16$

    $-2x^2 + 3xy - y^2 = -28$

12.4 Mixed Exercises

Solve each system.

31. $x^2 + 3y^2 = 3$

    $x = 3y$

32. $xy = 1$

    $x + 2y = 3$

33. $2x^2 + 3y^2 = 6$

    $x^2 + 3y^2 = 3$

34. $x^2 + 2xy - y^2 = 7$

    $x^2 \qquad - y^2 = 3$

35. $y = x^2$

    $y = 2x^2 - x - 6$

36. $3x^2 + \; y^2 = 13$

    $4x^2 - 3y^2 = 13$

37. $3x^2 + 2xy - 3y^2 = 5$

    $-x^2 - 3xy + \; y^2 = 3$

38. $x^2 - 3x + y^2 = 4$

    $2x - y \; = 3$

39. $xy = 5$

    $2x^2 - y^2 = 5$

40. $x^2 - xy + y^2 = 6$

    $x + y = 0$

12.5   Second-Degree Inequalities and Systems of Inequalities

Objective 1   Graph second-degree inequalities.

Graph each inequality.

1. $x \geq y^2$

2. $y^2 \geq 9 - x^2$

3. $16x^2 < 9y^2 + 144$

4. $25y^2 \leq 100 - 4x^2$

5. $x^2 + 4y^2 > 4$

6. $y \geq x^2 - 4$

7. $x \leq 2y^2 + 8y + 9$

8. $4y^2 \geq 196 + 49x^2$

Objective 2   Graph the solution set of a system of inequalities.

Graph each system of inequalities.

9. $-x + y > 2$
    $3x + y > 6$

10. $x + y > -2$
     $2x - y \leq -4$

11. $x - 2y \geq -6$
     $x + 4y \geq 12$

12. $x^2 + y^2 \leq 25$
     $3x - 5y > -15$

13. $9x^2 + 16y^2 < 144$
     $y^2 - x^2 > 4$

14. $x^2 + y^2 \leq 16$
     $y \leq x^2 - 4$

12.5 Mixed Exercises

Graph each inequality or system of inequalities.

15. $7x^2 \leq 42 - 6y^2$

16. $9x^2 + 64y^2 \leq 576$
     $x \geq 0$

17. $x^2 > 9 - y^2$
     $x \leq 0$
     $y \geq 0$

18. $x^2 - y^2 \leq 16$
     $y \geq 0$

19. $4y + x^2 < 0$
     $x \geq 0$

20. $x^2 + 4y^2 \leq 36$
     $-5 < x < 2$
     $y \geq 0$

ANSWERS TO
ADDITIONAL EXERCISES

Chapter R

PREALGEBRA REVIEW

R.1   Fractions

Objective 1

1. Prime
2. Composite
3. Neither
4. Prime
5. Prime
6. Composite
7. Composite
8. Prime
9. Prime
10. Composite
11. Composite
12. Prime

Objective 2

13. $2 \cdot 2 \cdot 17$
14. $3 \cdot 3 \cdot 5$
15. $2 \cdot 7 \cdot 7$
16. $2 \cdot 3 \cdot 13$
17. $2 \cdot 2 \cdot 2 \cdot 3$
18. $2 \cdot 3 \cdot 5 \cdot 7$
19. $3 \cdot 3 \cdot 3 \cdot 7$
20. $2 \cdot 2 \cdot 2 \cdot 2 \cdot 2 \cdot 2 \cdot 2 \cdot 2$
21. $2 \cdot 2 \cdot 2 \cdot 3 \cdot 7$
22. $2 \cdot 2 \cdot 2 \cdot 31$
23. 59
24. $2 \cdot 3 \cdot 7 \cdot 13$

Objective 3

25. $\frac{15}{28}$
26. $\frac{5}{7}$
27. $\frac{5}{14}$
28. $\frac{7}{2}$
29. $\frac{7}{25}$
30. $\frac{30}{19}$
31. $\frac{1}{5}$
32. $\frac{5}{6}$
33. $\frac{27}{52}$
34. $\frac{7}{5}$
35. $\frac{4}{9}$
36. $\frac{73}{33}$

Objective 4

37. $\frac{3}{4}$
38. $\frac{3}{22}$
39. 1
40. 12
41. $\frac{15}{2}$
42. $\frac{1}{6}$
43. $\frac{1}{2}$
44. $\frac{3}{4}$
45. $\frac{77}{2}$
46. $\frac{45}{4}$
47. $\frac{40}{3}$
48. $\frac{147}{8}$
49. $\frac{7}{5}$
50. $\frac{7}{4}$
51. 1
52. $\frac{3}{8}$
53. $\frac{2}{39}$
54. $\frac{1}{2}$
55. $\frac{1}{88}$
56. 2

57. $\frac{5}{7}$ **58.** $\frac{5}{3}$ **59.** $\frac{9}{4}$ **60.** 2

Objective 5

61. $\frac{4}{5}$ **62.** $\frac{2}{3}$ **63.** $\frac{2}{3}$ **64.** $\frac{29}{36}$

65. $\frac{2}{3}$ **66.** $\frac{256}{225}$ **67.** $\frac{27}{8}$ **68.** $\frac{83}{80}$

69. $\frac{25}{18}$ **70.** $\frac{125}{12}$ **71.** $\frac{81}{10}$ **72.** $\frac{331}{8}$

73. $\frac{1}{4}$ **74.** $\frac{5}{17}$ **75.** $\frac{2}{3}$ **76.** $\frac{5}{36}$

77. $\frac{7}{24}$ **78.** $\frac{11}{45}$ **79.** $\frac{1}{6}$ **80.** $\frac{25}{36}$

81. $\frac{31}{48}$ **82.** $\frac{49}{9}$ **83.** $\frac{7}{2}$ **84.** $\frac{1135}{16}$

R.1 Mixed Exercises

85. $2 \cdot 3 \cdot 5 \cdot 5$ **86.** $2 \cdot 2 \cdot 2 \cdot 2 \cdot 3 \cdot 7$ **87.** 47

88. $2 \cdot 2 \cdot 3 \cdot 5 \cdot 5 \cdot 5$ **89.** $2 \cdot 2 \cdot 3 \cdot 5 \cdot 7$ **90.** $2 \cdot 5 \cdot 31$

91. $\frac{22}{15}$ **92.** $\frac{4}{15}$ **93.** 6

94. $\frac{8}{21}$ **95.** $\frac{16}{13}$ **96.** $\frac{3}{4}$

97. $\frac{4}{7}$ **98.** $\frac{17}{90}$ **99.** $\frac{367}{60}$

100. $\frac{121}{24}$ **101.** 55   **102.** 2

R.2   Decimals and Percents

Objective 1

1. $\frac{3}{10}$ 2. $\frac{42}{100}$ 3. $\frac{26}{100}$ 4. $\frac{247}{1000}$

5. $\frac{234}{1000}$ 6. $\frac{427}{1000}$ 7. $\frac{7}{1000}$ 8. $\frac{92}{1000}$

9. $\frac{54}{10,000}$ 10. $\frac{186}{10}$ 11. $\frac{428}{100}$ 12. $\frac{8653}{1000}$

Objective 2

13. 39.48   14. 215.003   15. 480.8   16. 256.089

17. 29.514   18. 755.098   19. 16.35   20. 107.8

21. 15.05   22. 155.3   23. 65.33   24. 119.749

Objective 3

25. 40   26. 3236.1   27. 9.87   28. 2.3424

29. 4.122   30. 42.5691   31. 772   32. 635.333

33. 23.525   34. 40.363   35. 17.049   36. 24,250

37. 248   38. 4278.9   39. 467,040   40. 4.57

41. 3.948   42. .02687   43. 36,421   44. .429

45. 3840   46. .0074

Objective 4

47. .429   48. .3   49. .563   50. .625

51. $0.\overline{36}$, .364   52. $2.1\overline{6}$, 2.162   53. $.\overline{4}$; .444   54. 1.025

55. .235   56. $.0\overline{925}$; .093   57. .755   58. $.\overline{3}$; .333

Objective 5

59. .41   60. .29   61. .13   62. .004

63. .0028   64. .014   65. .275   66. 1

67. 3.62   68. 24%   69. 23%   70. 4.5%

71. 900%   **72.** 235%   **73.** .4%   **74.** 209%

75. 4.09%   **76.** .84%

R.2 Mixed Exercises

77. 644.661   **78.** 183.92   **79.** 20.88   **80.** 293.4

81. 377.89   **82.** 2178.64   **83.** 525.38   **84.** 2.387

85. 115.429   **86.** 73.238   **87.** 5200   **88.** .03504

89. $\frac{21}{25}$ **90.** .1875   **91.** .49   **92.** 4.2%

93. $.\overline{36}$; .364   **94.** $216.\overline{6}\%$ **95.** $.4\overline{6}$; .467   **96.** 245%

97. $\frac{27}{100}$ **98.** $266.\overline{6}\%$

Chapter 1

THE REAL NUMBER SYSTEM

1.1 Exponents, Order of Operations, and Inequality

Objective 1

1. 27

2. 32

3. 81

4. 16

5. $\frac{9}{16}$

6. $\frac{1}{64}$

7. $\frac{8}{125}$

8. $\frac{16}{81}$

9. .16

10. 0.000027

11. .81

12. 13.824

Objective 2

13. 17

14. 15

15. 9

16. −55

17. 45

18. 1

19. $\frac{7}{12}$

20. $\frac{41}{36}$

21. 2.44

22. $\frac{29}{2}$

23. $\frac{29}{16}$

24. −8

Objective 3

25. 102

26. 46

27. 52

28. 44

29. −2

30. 61

31. 68

32. −8

33. −71

34. 48

35. 176

36. 189

Objective 4

37. False

38. True

39. False

40. False

41. False

42. False

43. False

44. True

45. False

46. True

47. True

48. True

Objective 5

49. $7 = 13 - 5$

50. $9 > 16$

51. $12 \neq 14$

52. $22 > 17$

53. $9 + 13 > 21$

54. $19 < 35$

55. $6 \leq 6$

56. $30 - 7 > 20$

57. $7 > 15 \div 5$

58. $17 < 3 \cdot 10$

Objective 6

59. $12 > 9$

60. $8 \leq 12$

61. $4 > \frac{1}{10}$

62. $\frac{2}{3} > \frac{1}{2}$

63. $\frac{3}{4} > \frac{2}{3}$

64. $\frac{4}{5} \geq \frac{2}{7}$

65. $.19 < .21$

66. $.922 \geq .921$

67. $0 \leq 1$

68. $.01 < .1$

1.1 Mixed Exercises

69. $\frac{25}{9}$

70. 91.125

71. 52

72. 20

73. $\frac{1}{4}$

74. 96

75. True

76. False

77. $5(2+9) < 106$

78. $20 \geq 2(7)$

79. $.0002 < .002$

80. $\frac{3}{7} \geq \frac{3}{8}$

1.2 Variables, Expressions, and Equations

Objective 1

1. 5

2. 20

3. 0

4. 32

5. 28

6. 52

7. 0

8. $-\frac{2}{3}$

9. 0

10. $-\frac{16}{5}$

11. $\frac{28}{13}$

12. -3

Objective 2

13. $x+4$

14. $5x$

15. $3-x$

16. $2(x-4)$

17. $2x-7$

18. $3x-5$

19. $10x+21$ or $21+10x$

20. $4x-15$

21. $\dfrac{x+4}{2x}$

22. $\frac{2}{3}x-\frac{1}{2}x$

Objective 3

23. Yes

24. Yes

25. No

26. No

27. No

28. No

29. No

30. No

31. No

32. No

33. Yes

34. Yes

Objective 4

35. $x+6=10$

36. $x-5=9$

37. $5x+2=23$

38. $5+x=9$

39. $\frac{y}{9}=17$

40. $4x=2+3x$

41. $\frac{10}{x}=2+x$

42. $6x=18$

43. $\frac{x}{2}=0$

44. $\frac{x}{10}=1$

Objective 5

45. Expression

46. Equation

47. Equation

48. Expression

49. Expression

50. Equation

51. Expression

52. Equation

53. Equation

54. Expression

1.2 Mixed Exercises

55. -28

56. -2.4

57. -4

58. $-\frac{15}{2}$

59. **5**

60. $\frac{3}{8}$

61. No

62. Yes

63. No

64. Yes

65. No

66. Yes

67. $x + 4 = 10$

68. $6(5 + x) = 19$

69. $\frac{14}{x} = x + 2$

70. $\frac{24}{x} = x - 2$

1.3   Real Numbers and the Number Line

Objective 1

1. $-273; 103

2. $3500

3. −75

4. −2

5. $-72

6. 14,495

7. 279,867

8. $2\frac{1}{2}$

9. $-42,500

10. −396

11.

12.

13.

14.

15.

16.

17.

18.

19.

20.

Objective 2

21. −15

22. −2

23. −.820

24. −6.01

25. −2

26. $-\frac{1}{2}$

27. True

28. False

29. False

30. False

Objective 3

31. −23

32. 25

33. −4.5

34. $-\frac{3}{8}$

35. $\frac{5}{7}$

36. $-2\frac{3}{7}$

37. 0

38. −4

39. −22

Objective 4

40. 4

41. 143

42. 0

43. −95

44. −25

45. 2

46. −10

47. 7.52

48. −.9

49. $\frac{5}{6}$

50. $\frac{3}{4}$

51. $-2\frac{3}{8}$

52. $|2|$

53. $-|5|$

54. $-|10|$

1.3   Mixed Exercises

55. $2573

56. −27

57. 11,235

58. −800

59. 12.001

60. $-\frac{11}{12}$

61. $-\frac{2}{5}$

62. −4

63. $-|-3|$

64. $|-12|$

65. $-|-2.5|$

1.4 Adding Real Numbers

Objective 1

1. 15

2. 21

3. –5

4. –10

5. –18

6. –18

Objective 2

7. 2

8. –5

9. –7

10. $-\frac{1}{5}$

11. $-\frac{1}{6}$

12. $\frac{1}{35}$

13. $1\frac{3}{8}$

14. 0

15. –4.15 or –4.150

Objective 3

16. True

17. True

18. False

19. False

20. False

21. False

22. False

23. True

24. True

25. True

Objective 4

26. 1

27. –13

28. –23

29. –20

30. –16

31. –3

32. –6

33. –13.8

34. $-\frac{7}{8}$

35. $-\frac{73}{60}$

36. $\frac{3}{5}$

37. $\frac{3}{8}$

Objective 5

38. $-9+14;\ 5$

39. $-7+12;\ 5$

40. $-2+16;\ 14$

41. $-8+3+2;\ -3$

42. $\left[-4+(-3)\right]+10;\ 3$

43. $(-2+7)+9;\ 14$

44. $10+(-20+9);\ -1$

45. $-8+(-4)+(-11);\ -23$

46. $\left[-14+(-29)\right]+27;\ -16$

47. $\left[20+(-4)\right]+(-10);\ 6$

48. 8 yd lost

49. $495

50. 2654 ft

51. $16

52. 5°F

53. 260 ft

1.4 Mixed Exercises

54. –40

55. –1

56. –6

57. –7

58. –35

59. –6.8

60. $\frac{7}{10}$

61. $-11\frac{3}{8}$

62. $\left[4+(-12)\right]+6; \, -2$

63. $\left[25+(-20)\right]+(-2); \, 3$

64. $(-3)+(-4)+(-8); \, -15$

65. $\left[-2+(-4)\right]+20; \, 14$

1.5 Subtracting Real Numbers

Objective 1

1. 5

2. −3

3. 0

4. −12

5. −9

6. −6

Objective 2

7. −6

8. −4

9. −13

10. −11

11. 22

12. 8

13. 16

14. 4.4

15. 7.3

16. $\frac{3}{5}$

17. $-\frac{1}{30}$

18. $5\frac{7}{8}$

Objective 3

19. 7

20. .1

21. −6

22. 13

23. 0

24. 15

25. $\frac{4}{5}$

26. $-\frac{23}{18}$

27. −14

28. −2

Objective 4

29. $-9-3;\ -12$

30. $-4-(-13);\ 9$

31. $-6-(-2);\ -4$

32. $\left[-7-(-9)\right]-4;\ -2$

33. $-4-4;\ -8$

34. $-4-7;\ -11$

35. $-8-(-1-2);\ -5$

36. $-4-(-4-1);\ 1$

37. $\left[4+(-7)\right]-(-6);\ 3$

38. $\left[-4+(-2)\right]-(-12);\ 6$

39. −51.2°C

40. 3175 ft

41. $−263.16

42. 37°F

43. 1804 m

44. $-1147.32

1.5 Mixed Exercises

45. −10

46. 7

47. 46

48. −20

49. 0

50. −16.1

51. 10

52. 18

53. $-\frac{2}{3}$

54. $\frac{17}{16}$

55. $4-(-10); 14$

56. $(-4+12)-9; -1$

57. $\left[10-(-4)\right]-2; 12$

58. $\left[-4+(-8)\right]-2; -14$

1.6 Multiplying and Dividing Real Numbers

Objective 1

1. −28

2. −120

3. −72

4. −132

5. −320

6. −44

7. $-\frac{2}{15}$

8. $-\frac{7}{12}$

9. $-\frac{3}{13}$

10. −17.5

11. −36

12. −13.12

Objective 2

13. 12

14. 14

15. 182

16. 1000

17. 289

18. 40

19. $\frac{4}{5}$

20. $\frac{3}{4}$

21. $\frac{1}{6}$

22. 2.73

23. 2.84

24. 1.36

Objective 3

25. $\frac{1}{4}$

26. $-\frac{1}{2}$

27. −4

28. Does not exist

29. $\frac{5}{3}$

30. −3

31. $-\frac{12}{11}$

32. $\frac{24}{97}$

33. 4

34. −8

35. Does not exist

36. $\frac{8}{43}$

37. −8

38. −8

39. −4

40. 6

41. Undefined

42. 0

43. $-\frac{1}{6}$

44. $\frac{3}{8}$

45. $\frac{3}{7}$

46. $-\frac{2}{3}$

47. 5.2

48. −2.5

Objective 4

49. −11

50. 4

51. −46

52. −16

53. −4

54. −27

55. 64

56. 70

57. 9

58. $-\frac{7}{4}$

59. 2

60. 0

61. 5

62. –8

63. –6

64. $\frac{16}{21}$

65. $-\frac{3}{2}$

66. $\frac{5}{4}$

67. $\frac{37}{10}$

68. $\frac{41}{7}$

Objective 5

69. –5

70. 10

71. –96

72. –3

73. 34

74. –10

75. –3

76. 19

77. 80

78. $\frac{3}{4}$

79. $\frac{5}{16}$

80. $\frac{45}{4}$

81. $\frac{19}{4}$

82. $-\frac{5}{4}$

Objective 6

83. $4+(7)(-2);\ -10$

84. $-7+(-7)(3);\ -28$

85. $-2-(10)(-2);\ 18$

86. $-12-(-2)(7);\ 2$

87. $-2+2\big[14+(-4)\big];\ 18$

88. $85-\frac{3}{10}\big[50-(-10)\big];\ 67$

89. $.70\big[20+(-4)\big];\ 11.2$

90. $.85\big[32-(-4)\big];\ 30.6$

91. $\frac{2}{3}\big[16+(-10)\big]-(-34);\ 38$

92. $\frac{7}{8}\big[(-2)-6\big]+(-7);\ -14$

93. $\frac{-108}{-4};\ 27$

94. $\frac{50}{35+(-5)};\ \frac{5}{3}$

95. $-12+\frac{49}{-7};\ -19$

96. $9-\frac{-12}{4};\ 12$

97. $-3-\frac{-12}{-4};\ -6$

98. $\frac{(40)(-3)}{5-(-10)};\ -8$

99. $\frac{(-4)(7)}{-3+14};\ -\frac{28}{11}$

100. $\frac{14+(-4)}{-11-(-9)};\ -5$

Objective 7

101. $5x = -45$

102. $\frac{x}{-2} = -9$

103. $x + 9 = -8$

104. $x - (-7) = 12$

105. $\frac{2}{3}x = -7$

106. $x(-1) = 7$

107. $\frac{x}{-4} = 1$

108. $x - 9 = -4$

109. $-8x = 72$

1.6 Mixed Exercises

110. -48

111. 140

112. 42

113. -28

114. -16

115. 14

116. 35

117. 25

118. 17

119. -16

120. -30

121. -10

122. 17

123. 7

124. 1

125. -24

126. 0

127. Undefined

128. -1

129. $\frac{-11}{2}$

130. $\frac{7}{3}$

131. -4

132. $\frac{48}{23}$

133.   Undefined

134. 16

135. $\frac{100}{-16+(-9)} ; -4$

136. $\frac{(-40)(4)}{7-(-3)} ; -16$

137. $\frac{(-20)+(-10)}{3-(-3)} ; -5$

138. $\frac{-20}{-5} + \frac{100}{-25} ; 0$

1.7 Properties of Real Numbers

Objective 1

1. $y + 4 = 4 + y$

2. $5(2) = 2(5)$

3. $ab(2) = 2(ab)$

4. $7m = m \cdot 7$

5. $-4(p+9) = (p+9)(-4)$

6. $10\left(\frac{1}{4} \cdot 2\right) = \left(\frac{1}{4} \cdot 2\right)(10)$

7. $-4\left(\frac{1}{5}\right) = \frac{1}{5}(-4)$

8. $3 + (-4) = -4 + 3$

9. $2 + \left[10 + (-9)\right] = \left[10 + (-9)\right] + 2$

10. $-4(4+z) = (4+z)(-4)$

Objective 2

11. $x(9y) = \left[x(9)\right](y)$

12. $(4 \cdot 5)(-7) = 4\left[5(-7)\right]$

13. $\left[-4 + (-2)\right] + y = -4 + (-2 + y)$

14. $(2m)(-7) = (2)\left[m(-7)\right]$

15. $4(ab) = (4a)b$

16. $(-6x)(-2) = (-6)\left[x(-2)\right]$

17. $(-12x)(-y) = (-12)\left[x(-y)\right]$

18. $(-r)\left[(-p)(-q)\right] = \left[(-r)(-p)\right](-q)$

19. $\left[x + (-4)\right] + 3y = x + \left[(-4) + 3y\right]$

20. $4r + (3s + 14t) = (4r + 3s) + 14t$

Objective 3

21. 4　　　　22. -7　　　　23. -4　　　　24. 7　　　　25. $\frac{6}{7}$

Objective 4

26. $-4 + 4 = 0$; inverse

27. $-\frac{1}{7} + \frac{1}{7} = 0$; inverse

28. $1 \cdot 1 = 1$; either property

29. $\frac{2}{7} \cdot \frac{7}{2} = 1$; inverse

30. $-\frac{3}{5} \cdot -\frac{5}{3} = 1$; inverse

31. $\frac{8}{5}\left(\frac{5}{8}\right) = 1$; inverse

32. $-14 + 14 = 0$; inverse

33. $-9 + 9 = 0$; inverse

34. $0+0=0$; either property

35. $\left(-\frac{6}{17}\right)\left(-2\frac{5}{6}\right)=1$; inverse

36. $.25+(-.25)=0$; inverse

Objective 5

37. $y(6+7)$; $13y$

38. $r(10-4)$; $6r$

39. $az+2a$

40. $4(r+p)$

41. $3a+3b$

42. $4(c-d)$

43. $2an-4bn+6cn$

44. $7a-28b$

45. $-10y+18z$

46. $2k-7$

47. $-14(x+y)$

48. $2(7x+8z)$

1.7 Mixed Exercises

49. Associative

50. Commutative

51. Identity

52. Inverse

53. Associative

54. Distributive

55. Identity

56. Identity

57. Distributive

58. Commutative

59. $4y+9$

60. $-4r+3$

61. 11

62. –42

63. $-3x+2$

64. $4b+8$

1.8 Simplifying Expressions

Objective 1

1. $6+3y$

2. $12x-6y$

3. $8x+27$

4. $8n-14$

5. $-17+4b$

6. $7-d$

7. $10-6y$

8. $5+s$

9. $10x+3$

10. $-24p-10$

Objective 2

11. 4

12. –2

13. –7

14. 4

15. .3

16. –12

17. 125

18. 1

19. –16

20. $-\frac{3}{5}$

21. $-\frac{5}{9}$

22. $\frac{7}{9}$

Objective 3

23. Like

24. Unlike

25. Like

26. Unlike

27. Like

28. Like

29. Unlike

30. Like

31. Unlike

32. Unlike

33. Like

34. Unlike

Objective 4

35. $7a$

36. $-11x+10$

37. $5.9r+9.7$

38. $3a^3+2a^2$

39. $-\frac{1}{2}+\frac{1}{12}y$

40. $\frac{3}{10}r-\frac{1}{2}s$

41. $6x+10$

42. $12x-8$

43. $5r-4$

44. $-2t-6$

45. $5q+30$

46. $2a-16$

47. $3s-12$

48. $3x-4$

49. $-1.5y+16$

50. $-.2t-.2$

Objective 5

51. $7x+2x;\ 9x$

52. $3x-4x;\ -x$

53. $(6x+12)+4x;\ 10x+12$

54. $3(9+2x)+4x;\ 27+10x$

55. $3x-(7x+2);\ -4x-2$

56. $(5x-3)+4(x+2);\ 9x+5$

57. $(2-9x)-(10x+7);\ -19x-5$

58. $10-12(4-2x);\ -38+24x$

59. $4(2x-6x)+6(x+9);\ -10x+54$

60. $3(-7+5x)-4\big[2x-(-10)\big];\ -61+7x$

1.8   Mixed Exercises

61. $18a-12b$

62. $14x-5$

63. $35n-9$

64. $-10y^2+16y$

65. $1.7y^2-.5xy$

66. $2x-14$

67. $3s+42$

68. $-5x-11$

69. 93

70. –9

71. $\frac{1}{10}$

72. 5.6

73. Like

74. Unlike

75. Like

Chapter 2

SOLVING EQUATIONS AND INEQUALITIES

2.1 The Addition Property of Equality

Objective 1

1. Yes   **2.** No   **3.** No   **4.** No   **5.** No   **6.** Yes

Objective 2

7. 20   **8.** -1 **9.** -5 **10.** 1

11. $\frac{3}{2}$ **12.** $-\frac{3}{4}$ **13.** -5 **14.** $\frac{1}{2}$

15. -12.8 **16.** -2.4

Objective 3

17. -4 **18.** 3   **19.** 7   **20.** $\frac{1}{4}$

21. -8 **22.** -2 **23.** $\frac{5}{4}$ **24.** $-\frac{1}{3}$

25. 0   **26.** -5 **27.** $\frac{1}{3}$ **28.** 0

29. 7.2   **30.** 20

2.1 Mixed Exercises

31. -10 **32.** $-\frac{4}{3}$ **33.** -7

34. 4   **35.** $\frac{19}{36}$ **36.** $\frac{11}{8}$

37. -450

2.2 The Multiplication Property of Equality

Objective 1

1. 3	**2.** 9	**3.** −14	**4.** 3	**5.** 20
6. 6	**7.** −14	**8.** −42	**9.** −9	**10.** 7
11. −36	**12.** $\frac{8}{7}$	**13.** $\frac{4}{3}$	**14.** $\frac{7}{9}$	**15.** 6
16. 5	**17.** 4.3	**18.** 7.1	**19.** 5.8	**20.** 3.6
21. −8.2	**22.** −2.4	**23.** 6.4	**24.** −1.6	**25.** −2.7
26. 3.6				

Objective 2

27. 9	**28.** 8	**29.** 9	**30.** 4	**31.** −8
32. −7	**33.** 7	**34.** 8	**35.** 10	**36.** −3.9
37. 4	**38.** 26	**39.** $-\frac{7}{4}$	**40.** −8	**41.** 18
42. −5	**43.** 8	**44.** −8	**45.** −24	**46.** 7

2.2 Mixed Exercises

47. −8	**48.** −105	**49.** 6	**50.** $-\frac{10}{3}$	**51.** −1.5
52. −9.5	**53.** −6	**54.** 25	**55.** −26	**56.** 9.5
57. −7	**58.** $-\frac{3}{5}$			

2.3   More on Solving Linear Equations

Objective 1

1. $\frac{5}{2}$ 2. -7 3. 10 4. -5 5. -1

6. 2 7. $-\frac{1}{4}$ 8. $-\frac{5}{2}$.   9. 2 10. 4

11. -7 12. $-\frac{1}{5}$ 13. 1 14. 2

Objective 2

15. $-\frac{25}{2}$ 16. 2 17. $\frac{3}{5}$ 18. -14

19. $\frac{53}{11}$ 20. 4 21. $\frac{19}{4}$ 22. $-\frac{9}{10}$

23. 30 24. 40 25. -6 26. -5

27. -3 28. 10 29. $40,000$ 30. 6875

Objective 3

31. No solution   32. No solution   33. All real numbers

34. No solution   35. No solution   36. All real numbers

37. No solution   38. No solution   39. No solution

40. No solution   41. No solution   42. All real numbers

43. No solution   44. No solution

Objective 4

45. $36 - m$ 46. $\dfrac{17}{p}$ 47. $10q$ cents

2.3   Mixed Exercises

48. -5 49. -5 50. 3

51. $\frac{5}{2}$ 52. 0 53. -4

54. $\frac{1}{3}$ 55. $-\frac{7}{2}$ 56. 0

57. 3 58. No solution   59. No solution

60. All real numbers   61. -35 62. 6

63. All real numbers   **64.** $-\frac{3}{2}$ **65.** $6x + 4y$

2.4   An Introduction to Applied Problems

Objectives 1, 2, 3

1. $4 + 3x = 7$; 1

2. $4x - 2 = 3 + 6x$; $-\frac{5}{2}$

3. $-2(4 - x) = 24$; 16

4. $6(x - 4) = x(-2)$; 3

5. $-3(x - 4) = -5x + 2$; -5

6. $7 + 4x = 6x - 5$; 6

7. 27 inches

8. 209 votes

9. 52

10. Mount McKinley: 20,320 ft,
Mount Rainier: 14,410 ft

11. $4.25

12. 16 oz of cranberry juice;
32 oz of orange juice;
128 oz of ginger ale

13. Mark 3; Pablo 14; Faustino 12

14. 8 ft

Objective 4

15. $133°$ **16.** $20°$ **17.** $66°$ **18.** $49°$ **19.** $27°$

20. $45°$ **21.** $76°$ **22.** $55°$ **23.** $55°$ **24.** $43°$

Objective 5

25. 76, 78   **26.** 96, 98   **27.** 27, 28   **28.** 13, 15   **29.** 13, 15

2.4   Mixed Exercises

30. 27

31. 76 men

32. 22 cm; 48 cm; 60 cm

33. $59°$ or $31°$

34. 119, 121, and 123

35. $60°$

36. 21

37. 75, 76, 77

38. $78°$

39. $420

2.5 Formulas and Geometry Applications

Objective 1

1. 24	**2.** 15	**3.** 10	**4.** 7	**5.** 7
6. 28.26	**7.** 2400	**8.** 40	**9.** 95	**10.** 12
11. 113.04	**12.** 100.48			

Objective 2

13. 36 ft   **14.** 8 inches   **15.** 12 ft   **16.** 200.96 sq in

17. 31,400 sq ft   **18.** 12 m   **19.** 50 sq ft   **20.** 3052.08 cu cm

21. $1\frac{1}{2}$ years   **22.** 4 cm

Objective 3

23. $(3x)^{o} = 45^{o};\ (9x)^{o} = 135^{o}$

24. $(3x+5)^{o} = 35^{o};\ (6x-25)^{o} = 35^{o}$

25. $(3x-30)^{o} = 120^{o};\ (x+10)^{o} = 60^{o}$

26. $(6x)^{o} = 72^{o};\ (10x-48)^{o} = 72^{o}$

27. $(7x+3)^{o} = 150^{o};\ (8x-18)^{o} = 150^{o}$

28. $(2x+16)^{o} = 48^{o};\ (7x+20)^{o} = 132^{o}$

29. $(4x+7)^{o} = 59^{o};\ (11x-22)^{o} = 121^{o}$

30. $(9x)^{o} = 126^{o};\ (3x+12)^{o} = 54^{o}$

31. $(4x-11)^{o} = 129^{o};\ (x+16)^{o} = 51^{o}$

32. $(23x-13)^{o} = 148^{o};\ (19x+15)^{o} = 148^{o}$

Objective 4

33. $H = \dfrac{V}{LW}$

34. $p = \dfrac{A}{1+rt}$

35. $r = \dfrac{S-a}{S}$

36. $h = \dfrac{2A}{b}$

37. $h = \dfrac{S-2\pi r^{2}}{2\pi r} = \dfrac{S}{2\pi r} - r$

38. $n = \dfrac{a_n - a_1 + d}{d}$

39. $A = \dfrac{P}{1-rt}$

40. $h = \dfrac{V}{\pi r^{2}}$

41. $b = \dfrac{2A - Bh}{h}$

42. $a_1 = \dfrac{2S_n - a_n n}{n}$

43. $F = \frac{9}{5}C + 32$

44. $v = \dfrac{d - gt^{2}}{t}$

45. $h = \dfrac{3V}{\pi r^2}$

46. $n = \dfrac{S+360}{180}$

2.5 Mixed Exercises

47. $L = \dfrac{P-2W}{2}$; 111

48. $H = \dfrac{V}{LW}$; 5

49. $h = \dfrac{3V}{\pi r^2}$; 3

50. $b = \dfrac{2A-Bh}{h}$; 3

51. $a = S(1-r)$; 36

52. $r = \dfrac{A-P}{At}$; .04

53. 267.95 cu in

54. 550 m

55. 14 cm

56. 970 ft

57. 35 m

58. 75.36 sq cm

59. $(5x-20)^o = 65^o$; $(3x+14)^o = 65^o$

60. $(3x+42)^o = 81^o$; $(10x-31)^o = 99^o$

2.6 Ratio and Proportion Percent

Objective 1

1. $\frac{8}{3}$ 2. $\frac{7}{9}$ 3. $\frac{5}{7}$ 4. $\frac{16}{1}$ or 16

5. $\frac{3}{4}$ 6. $\frac{3}{16}$ 7. $\frac{5}{36}$ 8. $\frac{69}{10}$

9. $\frac{5}{24}$ 10. $\frac{5}{63}$ 11. $\frac{3}{16}$ 12. $\frac{1}{20}$

13. $\frac{1}{5}$ 14. $\frac{7}{8}$ 15. 5-lb box   16. 64-oz size

17. 20-count box   18. 48-oz bottle   19. 48-oz jar   20. 7-oz size

Objective 2

21. 15   22. $\frac{8}{5}$ 23. 12   24. 4

25. $\frac{25}{12}$ 26. $\frac{10}{3}$ 27. 14   28. $-\frac{41}{3}$

29. $\frac{20}{19}$ 30. -13 31. $-\frac{7}{4}$ 32. $\frac{44}{5}$

Objective 3

33. 75 min   34. $27   35. 15 inches   36. 43.2 acres

37. 4.5 oz   38. 14 tanks   39. $5.75   40. 10 lb

41. $135   42. $250

Objective 4

43. 189   44. 90   45. 87.5   46. 7.25%

47. 600%   48. 2%   49. 324   50. 750

51. 14.95   52. 17%   53. 58 females   54. $7950

55. 4.25%   56. 70%

2.6 Mixed Exercises

57. $\frac{20}{7}$ 58. $\frac{1}{32}$ 59. $\frac{1}{8}$ 60. $\frac{5}{28}$

61. $\frac{7}{8}$ 62. $\frac{5}{4}$ 63. 96-oz carton   64. 42-oz size

65. 1   66. 24   67. $\frac{16}{7}$ 68. $\frac{7}{2}$

69. $\frac{4}{5}$ **70.** 4   **71.** $20.25; $24.75   **72.** 22%

73. $112.50   **74.** $15

2.7 The Addition and Multiplication Properties of Inequality

Objective 1

1.

2.

3.

4.

5.

6.

7.

8.

9.

10.

11.

12.

Objective 2

13. $j \le 5$

14. $m > -4$

15. $b \ge 1$

16. $t > 3$

17. $y > -5$

18. $a \ge 3$

19. $a \le -4$

20. $x > 2$

21. $p \le 0$

22. $b < -2$

Objective 3

23. $x \le 5$

24. $s < -2$

25. $q \le -5$

26. $r > 10$

27. $r > -4$

28. $k \ge -4$

29. $k \geq 0$

30. $n \geq 0$

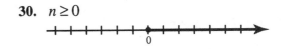

31. $z < -6$

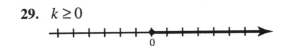

32. $t \geq -5$

33. $t \geq 7$

34. $m < 4$

Objective 4

35. $(1, 9)$

36. $(-8, 6)$

37. $(5, 9)$

38. $[-9, -7)$

39. $(4, 5)$

40. $\left(-\dfrac{8}{3}, 0\right)$

41. $m \geq 4$

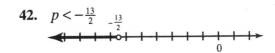

42. $p < -\dfrac{13}{2}$

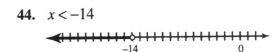

43. $p < -1$

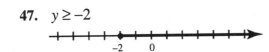

44. $x < -14$

45. $y < \dfrac{8}{3}$

46. $z \geq \dfrac{8}{5}$

47. $y \geq -2$

48. $z \geq 3$

Objective 5

49. $315 or more

50. All numbers less than or equal to 6

51. 89 or more

52. 81 or more

53. $38.84 or less

54. $7.75 or more

55. 55 or more

56. 25 or more

57. 10 ft

58. 17 cm

2.7 Mixed Exercises

59. $y \geq 7$

60. $k \geq -4$

61. $x \leq 3$

62. $k \geq 0$

63. $b \leq -3$

64. $m > -\frac{15}{2}$

65. $p > -3$

66. $\frac{9}{5} \geq w$

67. $x \leq 2.5$

68. $x < 1$

69. $x < -1$

70. $x \leq 8$

71. $y \leq 0$

72. $t < -11$

73. $r > 7$

74. $x \leq 8$

75. All numbers less than 5

76. At least 9 cars

77. All numbers greater than 5

78. 47 cm

79. $565 or more

80. 53 meters or less

Chapter 3

LINEAR EQUATIONS AND INEQUALITIES IN TWO VARIABLES

3.1 Reading Graphs; Linear Equations in Two Variables

Objective 1

1. 45%

2. $11,600

3. $12,760

4. 1987

5. 1990 – 1991

6. 1988 – 1989

In Exercises 7 and 8, percents may vary slightly from those given due to difficulty in reading graph precisely.

7. 5.5%; Quarter 1, 1989

8. 1.7%; Quarter 4, 1992

9. Quarter 1, 1991

10. 1991 – 1992

11. 280

12. 40

13. 1993 – 1994

14. 280

15. 27

16. Wednesday

17. 884

18. 81

19. 2655

20. Friday

21. 450

22. U.S.

23. Japan

24. Germany

Objective 2

25. $(4,7)$

26. $(-7,2)$

27. $(-4,0)$

28. $(2,7)$

29. $\left(\frac{1}{3},-9\right)$

30. $(-2,-3)$

31. $(4,7)$

32. $\left(0,\frac{1}{3}\right)$

33. $(.2,.3)$

34. $(9,-7)$

Objective 3

35. Yes

36. No

37. No

38. No

39. Yes

40. Yes

41. Yes

42. Yes

43. Yes

44. No

Objective 4

45. **(a)** $(2,-1)$ **(b)** $(0,-5)$ **(c)** $(4,3)$ **(d)** $(-1,-7)$ **(e)** $(7,9)$

46. **(a)** $(1,2)$ **(b)** $(-4,12)$ **(c)** $(-1,6)$ **(d)** $(-6,16)$ **(e)** $(5,-6)$

47. **(a)** $(-4,-5)$ **(b)** $(2,7)$ **(c)** $\left(-\frac{3}{2},0\right)$ **(d)** $(-2,-1)$ **(e)** $(-5,-7)$

48. **(a)** $(2,0)$ **(b)** $\left(4,-\frac{5}{2}\right)$ **(c)** $\left(-\frac{2}{5},3\right)$ **(d)** $\left(0,\frac{5}{2}\right)$ **(e)** $(2,0)$

49. **(a)** $(-2,-2)$ **(b)** $(-2,0)$ **(c)** $(-2,19)$ **(d)** $(-2,3)$ **(e)** $\left(-2,-\frac{2}{3}\right)$

50. **(a)** $(2,4)$ **(b)** $(0,4)$ **(c)** $(4,4)$ **(d)** $(-4,4)$ **(e)** $(.75,4)$

Objective 5

51.

x	2	$\frac{1}{2}$	1
y	-2	4	2

52.

x	2	-2	0
y	3	-3	0

53.

x	0	$\frac{4}{3}$	4
y	2	0	-4

54.

x	0	3	6
y	-4	2	8

55.

x	0	3	-1
y	9	0	12

56.

x	0	4	-8
y	6	0	18

57.

x	-2	-2	-2
y	0	4	-5

58.

x	-5	0	7
y	2	2	2

59.

x	4	4	4
y	-8	4	8

60.

x	-6	0	6
y	4	4	4

61.

x	y
0	$\frac{3}{2}$
-2	0
2	3

62.

x	y
0	4
3	0
$\frac{15}{4}$	-1

63.

x	y
2	0
0	-7
3	$\frac{7}{2}$

64.

x	y
0	3
4	0
8	-3

65.

x	y
-4	4
0	4
6	4

66.

x	y
1	-3
1	0
1	5

Objective 6

67. – 78.

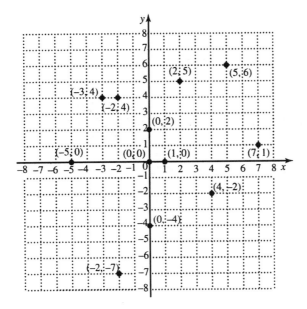

79. IV   **80.** IV

81. III   **82.** IV

83. I   **84.** None

85. None   **86.** None

3.1 Mixed Exercises

87. 1985 – 1990   **88.** 10,440   **89.** 3360

90. 1980 – 1985   **91.** 20,880   **92.** 2520

93. No   **94.** No   **95.** Yes

96. No   **97.** Yes   **98.** No

99. $(6,-10)$, $\left(-\frac{1}{2},3\right)$, $(1,0)$ **100.** $(0,3)$, $(4,0)$, $\left(2,\frac{3}{2}\right)$

101. $(3,-2)$, $\left(-\frac{1}{2},5\right)$, $(2.5,-1)$ **102.** $(0,-4)$, $(-1,-4)$, $(100,-4)$

103.

x	3	–2	–4
y	10	0	–4

104.

x	0	$\frac{5}{4}$	1
y	$\frac{5}{3}$	0	$\frac{1}{3}$

105.

x	–7	–7	–7
y	–5	0	10

106.

x	2	0	4
y	0	5	–5

107.

x	y
0	5
6	0
3	$\frac{5}{2}$

108.

x	y
–10	0
0	5
–4	3

3.2 Graphing Linear Equations in Two Variables

Objective 1

1. $(0,3)$, $(3,0)$, $(2,1)$

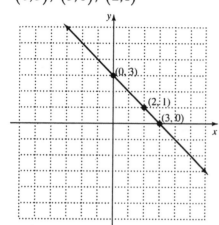

2. $(0,7)$, $(-7,0)$, $(-4,3)$

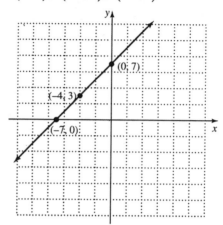

3. $(0,-3)$, $(4,-3)$, $(-3,-3)$

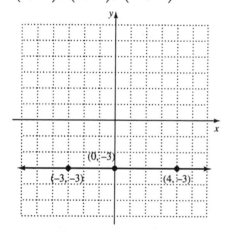

4. $(0,-2)$, $(-2,0)$, $(3,-5)$

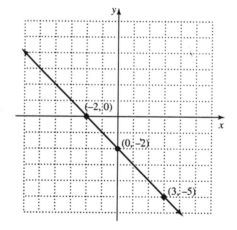

5. $(0,2)$, $(-4,0)$, $(-2,1)$

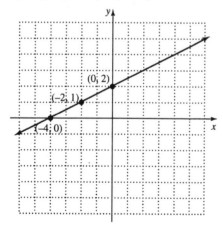

6. $(4,0)$, $(4,-2)$, $(4,3)$

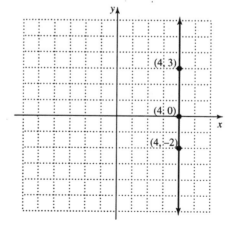

7. $(0,-4)$, $(4,0)$, $(-2,-6)$

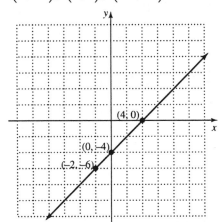

8. $(0,-2)$, $\left(\frac{2}{3},0\right)$, $(2,4)$

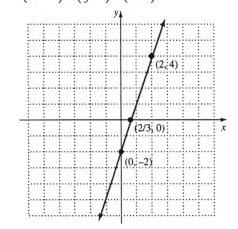

9. $(0,1)$, $(2,0)$, $(-2,2)$

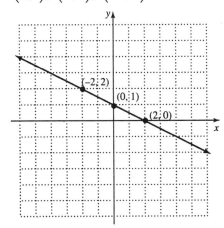

10. $(0,1)$, $(-1,0)$, $(4,5)$

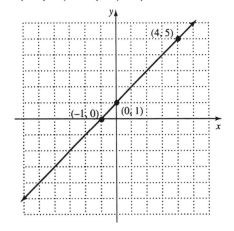

11. $(0,-3)$, $\left(\frac{3}{2},0\right)$, $(1,-1)$

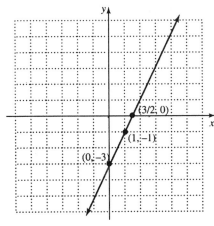

12. $(0,2)$, $(3,0)$, $(-3,4)$

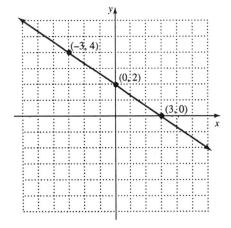

Objective 2

13. x-intercept: $(-2,0)$;
y-intercept: $(0,5)$

14. x-intercept: $(4,0)$;
y-intercept: $(0,6)$

15. x-intercept: $(0,0)$;
y-intercept: $(0,0)$

16. x-intercept: $(2,0)$;   **17.** x-intercept: $(2,0)$;   **18.** x-intercept: $\left(\frac{9}{4},0\right)$;
    y-intercept: $\left(0,\frac{8}{5}\right)$ y-intercept: $(0,-5)$ y-intercept: $(0,3)$

19. x-intercept: $\left(-\frac{2}{3},0\right)$;   **20.** x-intercept: $\left(\frac{12}{5},0\right)$;   **21.** x-intercept: $\left(-\frac{9}{2},0\right)$;
    y-intercept: $(0,-1)$ y-intercept: $(0,-4)$ y-intercept: $(0,-1)$

22. x-intercept: $(3,0)$;
    y-intercept: $\left(0,\frac{9}{4}\right)$

23. x-intercept: $(2,0)$ **24.** x-intercept: $\left(\frac{5}{2},0\right)$
    y-intercept: $(0,6)$ y-intercept: $(0,3)$

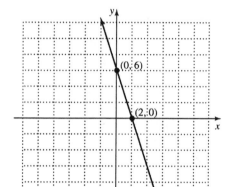

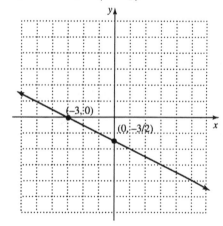 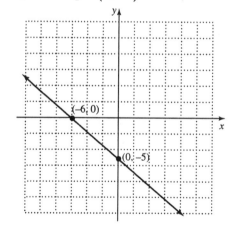

25. x-intercept: $(-3,0)$ **26.** x-intercept: $(-6,0)$
    y-intercept: $\left(0,-\frac{3}{2}\right)$ y-intercept: $(0,-5)$

27. x-intercept: $(1,0)$

y-intercept: $(0,-4)$

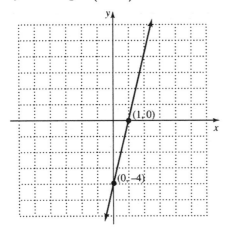

28. x-intercept: $(3,0)$

y-intercept: $(0,-2)$

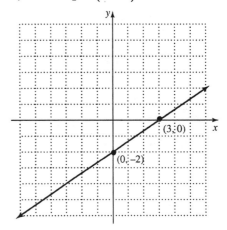

29. x-intercept: $(-4,0)$

y-intercept: $(0,-3)$

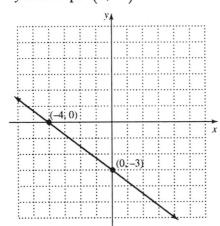

30. x-intercept: $(4,0)$

y-intercept: $(0,2)$

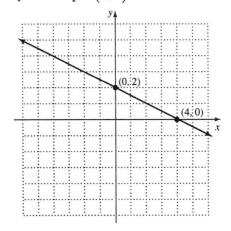

31. x-intercept: $(3,0)$

y-intercept: $(0,-2)$

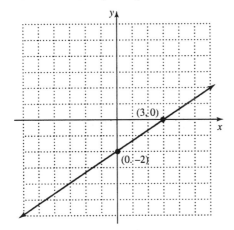

32. x-intercept: $(-2,0)$

y-intercept: $(0,5)$

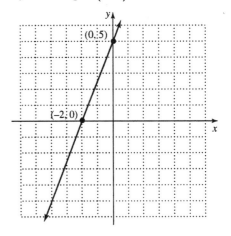

Objective 3

33.

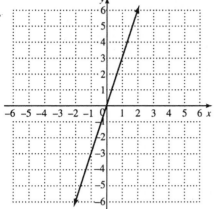

34.

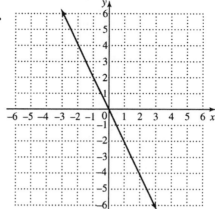

35.

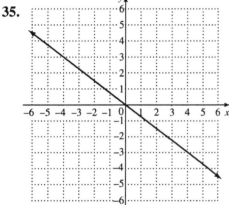

36.

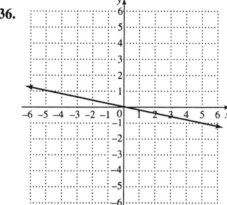

37.

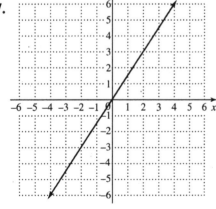

38.

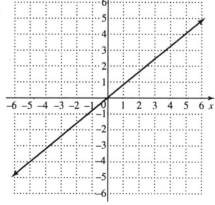

39.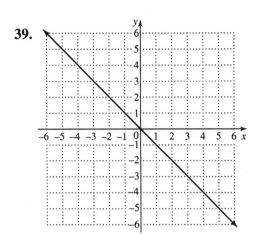

40.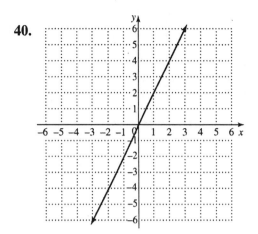

41.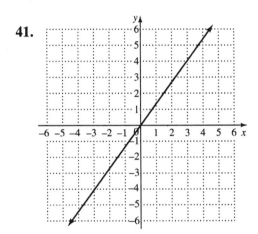

42.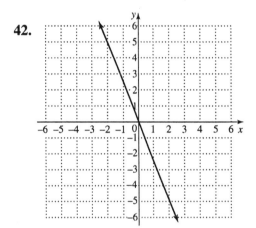

43.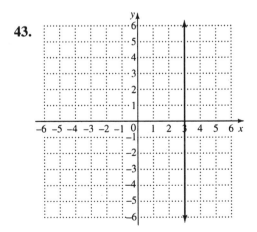

44.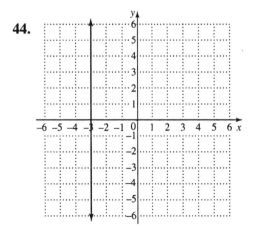

45.

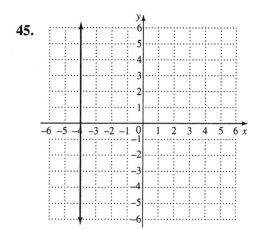

46.

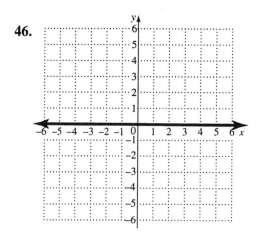

47.

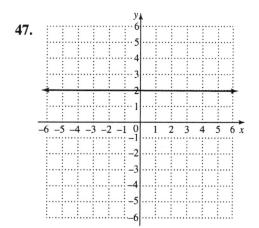

48.

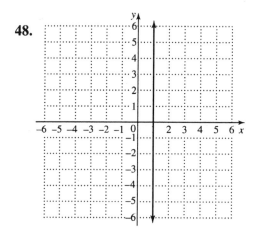

49.

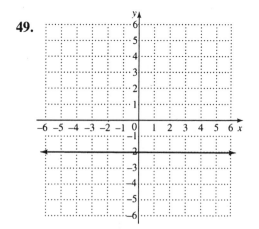

50.

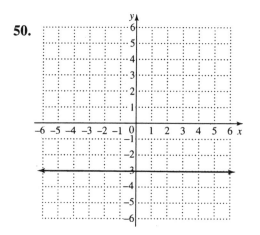

51.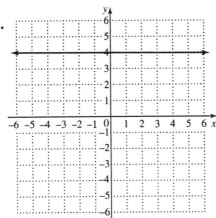

52.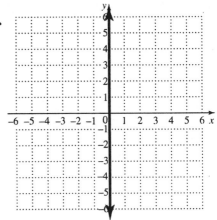

Objective 5

53. 1990: 2435; 1991: 2350; 1992: 2265; 1993: 2180; 1994: 2095; 1995: 2010

54. 1994: $4.9 million;   1995: $5.53 million;   1996: $6.16 million   1997: $6.79 million

55. 1993: 325;   1994: 367;   1995: 409;   1996: 451

56. (a) $45   (b) $42   **(c)** $33   **(d)** $18

3.2 Mixed Exercises

57. $(0,4)$, $(10,0)$, $(5,2)$

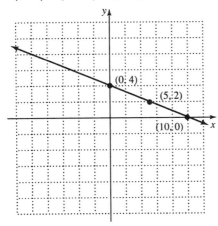

58. $(0,2)$, $(-6,0)$, $(6,4)$

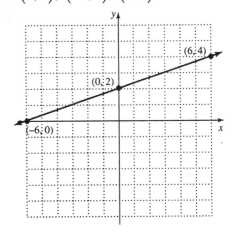

59. $(3,1), (-1,0), (-5,-1)$

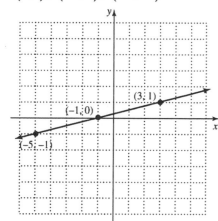

60. $(0,4), (-2,0), (-4,-4)$

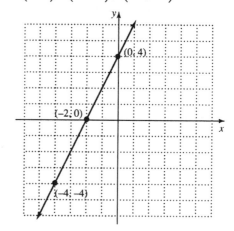

61. $\left(0,-\frac{1}{2}\right), (1,0), (-3,-2)$

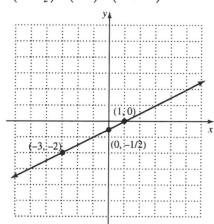

62. $(0,3), (-3,0), (1,4)$

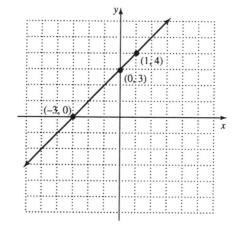

63. x-intercept: $(4,0)$

   y-intercept: $(0,6)$

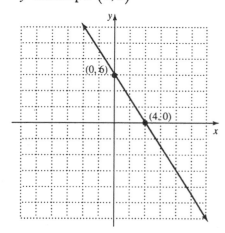

64. x-intercept: $\left(\frac{8}{3},0\right)$

   y-intercept: $(0,-4)$

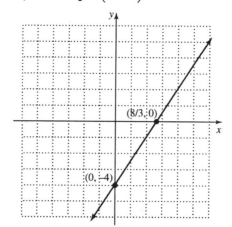

65. x-intercept: $(-5,0)$

y-intercept: $(0,-2)$

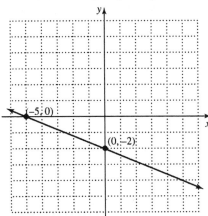

66. x-intercept: $(6,0)$

y-intercept: None

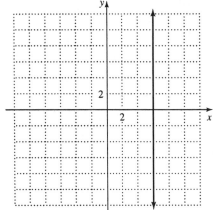

67. x-intercept: None

y-intercept: $(0,-5)$

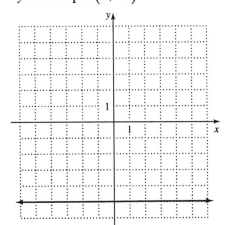

68. x-intercept: $(-2,0)$

y-intercept: $\left(0,\frac{8}{7}\right)$

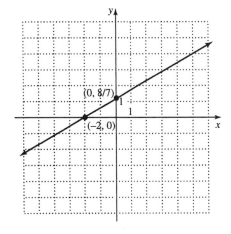

69. x-intercept: $(0,0)$

y-intercept: $(0,0)$

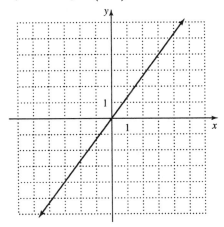

70. x-intercept: $(0,0)$

y-intercept: $(0,0)$

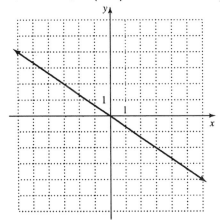

71. x-intercept: $\left(\frac{3}{2},0\right)$

   y-intercept: $(0,2)$

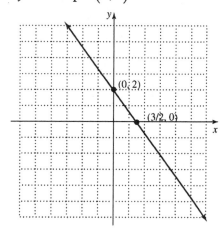

72. x-intercept: $(0,0)$

   y-intercept: $(0,0)$

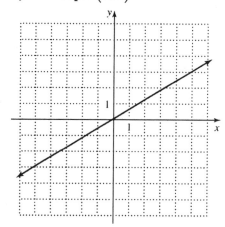

3.3 Slope of a Line

Objective 1

1. -2
2. 1
3. $\frac{1}{5}$
4. -3
5. $-\frac{3}{5}$

6. $-\frac{1}{14}$
7. 1
8. 0
9. 1
10. $-\frac{7}{6}$

11. Undefined slope
12. -2

Objective 2

13. -5
14. $\frac{1}{2}$
15. $-\frac{2}{5}$
16. $-\frac{4}{7}$
17. $\frac{3}{4}$

18. $\frac{2}{3}$
19. $-\frac{2}{7}$
20. $\frac{4}{7}$
21. $\frac{4}{3}$
22. 0

23. Undefined slope
24. $\frac{3}{4}$

Objective 3

25. -5; 5; neither
26. 4; $-\frac{1}{4}$; perpendicular

27. 1; 1; parallel
28. -1; 1; perpendicular

29. -2; $-\frac{1}{4}$; neither
30. -3; $\frac{1}{3}$; perpendicular

31. -2; $-\frac{5}{3}$; neither
32. -4; -1; neither

33. 0; 0; parallel
34. 0; undefined slope; perpendicular

35. $\frac{4}{3}$; $\frac{4}{3}$; parallel
36. $-\frac{6}{5}$; $\frac{6}{5}$; neither

Objective 4

37. $1250/yr
38. 1500

39. (a) $7540/yr
40. 113.5 ft/min

   (b) $6310/yr

   (c) $6925/yr

41. .24 horizontal ft/1 vertical ft
42. (a) -350 students/yr

   (b) The enrollment was decreasing.

3.3 Mixed Exercises

43. 2

44. 1

45. −2

46. 0

47. Undefined slope

48. $-\frac{2}{5}$

49. $\frac{4}{3}$

50. $\frac{2}{3}$

51. 3

52. Undefined slope

53. $\frac{1}{7}$

54. 2

55. −1; $\frac{5}{4}$; neither

56. $-\frac{3}{2}$; $\frac{1}{2}$; neither

57. Undefined slope; 0; perpendicular

58. 0; 0; parallel

59. $\frac{8}{9}$; $\frac{8}{9}$; parallel

60. $\frac{3}{11}$; $-\frac{3}{11}$; neither

61. 5.4 employees/yr

62. 3,033,100

3.4   Equations of a Line

Objective 1

1. $2x - y = 5$

2. $6x - y = 2$

3. $4x + y = 3$

4. $5x + y = 3$

5. $2x + 3y = 6$

6. $x + 4y = -12$

7. $3x - 5y = -2$

8. $6x - 5y = 1$

9. $7x - 3y = -27$

10. $y = 3$

Objective 3

11. $2x - y = -9$

12. $4x - y = 2$

13. $5x - y = 21$

14. $x - y = -9$

15. $5x + y = -1$

16. $3x + y = 3$

17. $x + 2y = 1$

18. $2x + 3y = -13$

19. $3x + 4y = -15$

20. $4x + 5y = 2$

21. $x = 3$

22. $y = 2$

23. $y = -4$

24. $x = 0$ (the y-axis)

25. $x = 2$

26. $y = 0$ (the x-axis)

Objective 4

27. $x - y = -5$

28. $4x + y = 29$

29. $3x + 2y = 23$

30. $x + 3y = -1$

31. $x + 2y = -2$

32. $7x + 4y = 13$

33. $x + 2y = -7$

34. $x + y = -5$

35. $y = 1$

36. $y = -5$

37. $x = 0$ (the y-axis)

38. $x = -1$

Objective 5

39. $x - y = 11$

40. $2x + 3y = 9$

41. $x + 3y = -5$

42. $4x - 3y = -17$

43. $5x + y = 4$

44. $3x + y = -28$

45. $x - 5y = 7$

46. $2x + 3y = 1$

47. $y = 7$

48. $x = 2$

49. $y = 6$

50. $x = -3$

Objective 6

51. (a) $y = .13x$

    (b) $(0, 0), (5, .65), (10, 1.30)$

52. (a) $y = 8x$

    (b) $(0, 0), (5, 40), (10, 80)$

53. (a) $y = 1.25x$

    (b) $(0, 0), (5, 6.25), (10, 12.50)$

54. (a) $y = .13x + .35$

    (b) $(0, .35), (5, 1.00), (10, 1.65)$

55. (a) $y = 8x + 15$

    (b) $(0, 15), (5, 55), (10, 95)$

56. (a) $y = 1.25x + 25$

    (b) $(0, 25), (5, 31.25), (10, 37.50)$

57. $1.91 = .13x + .35$; 12 min

58. $207 = 8x + 15$; 24 rows

59. $62.50 = 1.25x + 25$; 30 lines

3.4 Mixed Exercises

60. $15x + 24y = -16$

61. $x - y = -3$

62. $4x + 7y = -56$

63. $x = -3$

64. $3x + 4y = -8$

65. $y = -5$

66. $y = 2$

67. $x = -2$

68. $x - y = -3$

69. $3x + y = 25$

70. $3x + y = -1$

71. $y = 5$

72. $-\dfrac{2}{7}$; $(0, 2)$

73. $\dfrac{3}{2}$; $\left(0, -\dfrac{9}{2}\right)$

74. undefined slope; no y-intercept

75. 0; $(0, 2)$

3.5   Graphing Linear Inequalities in Two Variables

Objective 1

1.
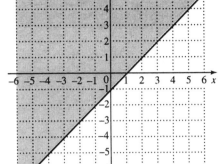

2.
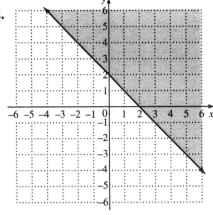

3.
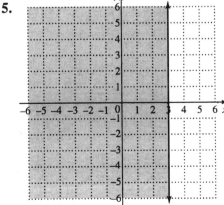

4.
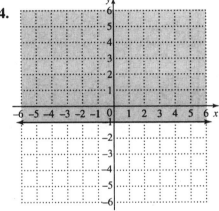

5.
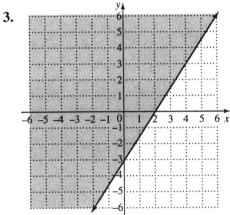

6.
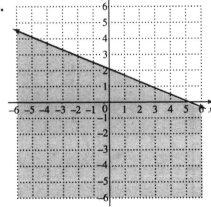

7.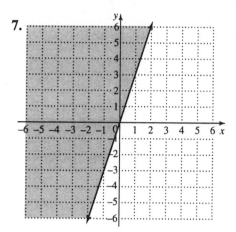

8.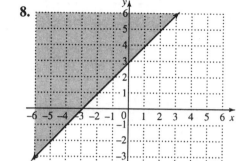

9.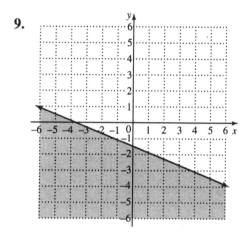

10.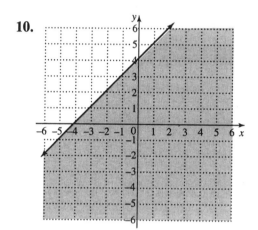

11.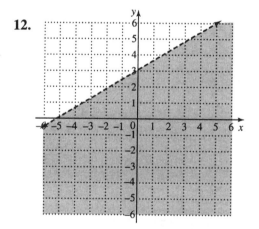

12.

13.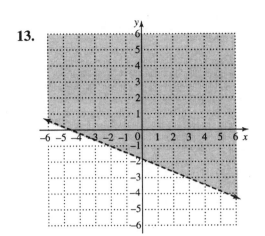

14.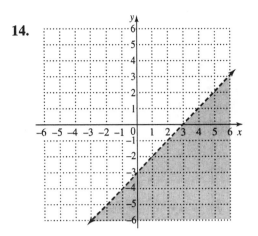

15.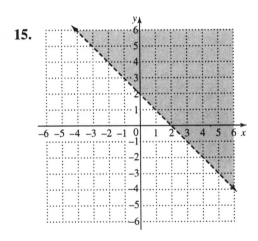

16.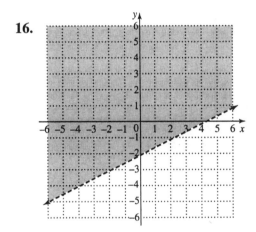

17.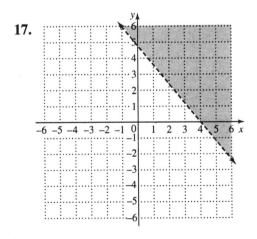

18.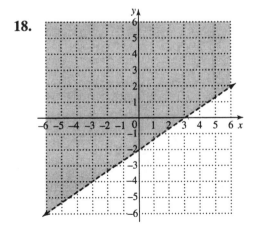

19.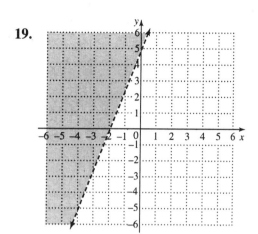

20.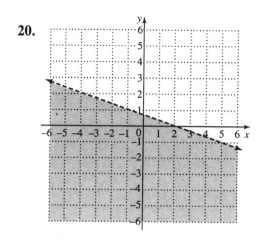

Objective 2

21.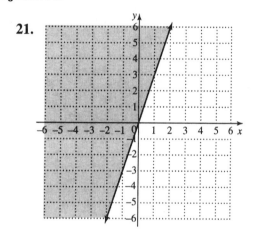

22.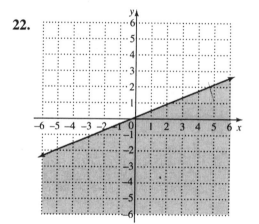

23.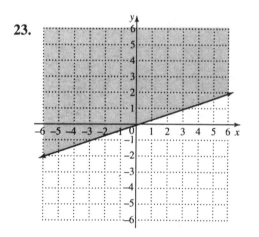

24.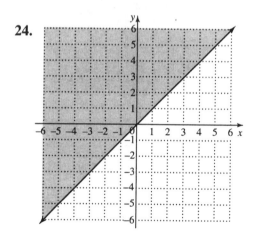

25.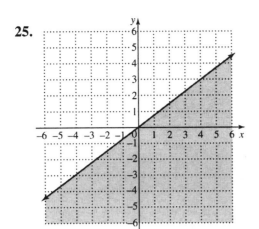

26.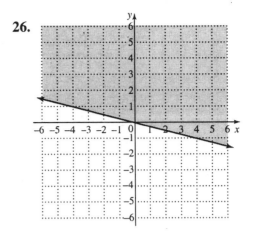

27.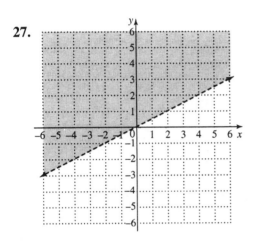

28.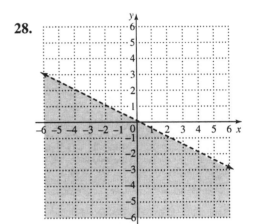

29.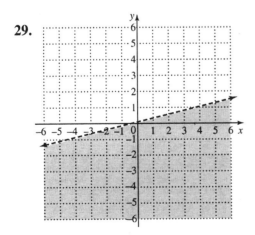

30.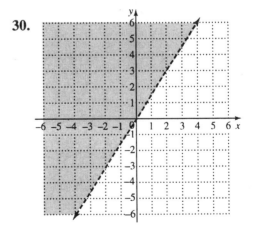

3.5 Mixed Exercises

31.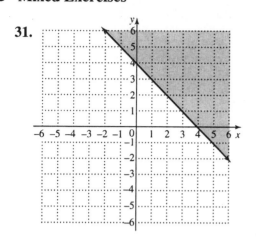

32.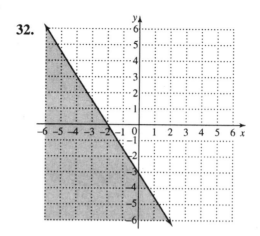

33.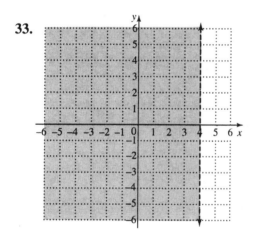

34.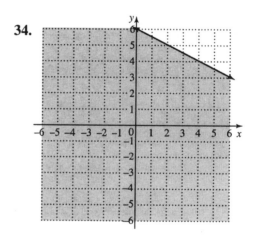

35.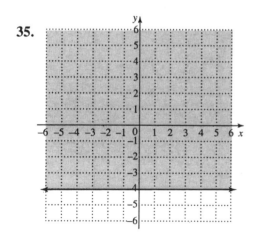

36.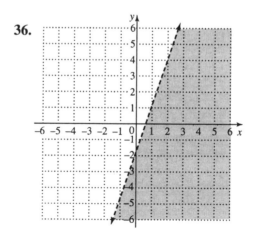

37.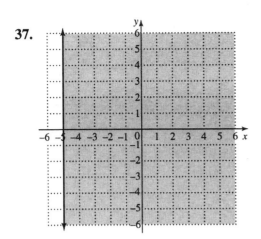

38.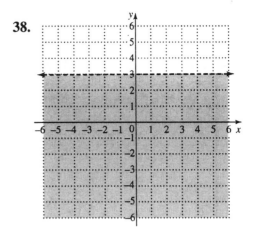

39.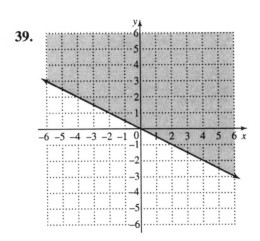

40.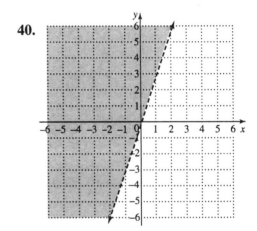

41.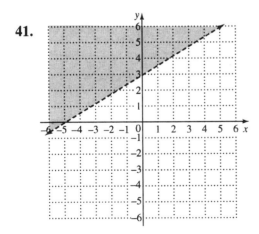

42.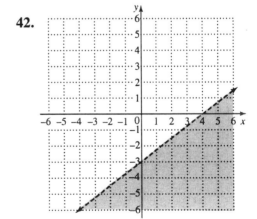

3.6 Introduction to Functions

Objective 1

1. not a function

2. function

3. function

4. function

5. not a function

6. function

7. not a function

8. not a function

9. function

10. function

11. not a function

12. not a function

13. function

14. function

15. function

Objective 2

16. function; domain: $\{-1, -2, 0\}$; range: $\{1, 2, 0\}$

17. function; domain: $\{3, 2, 1, -1\}$; range: $\{0, 4, 6, 3\}$

18. function; domain: $\{-2, -1, 0, 1\}$; range: $\{-2, -1, 0\}$

19. function; domain: $\{3, 2, 1\}$; range: $\{5, 3, 0\}$

20. not a function; domain: $\{1, 2, -1\}$; range: $\{3, -1, 4\}$

21. function; domain: $\{2, 1, -1, 0\}$; range: $\{-4, -2, 2, 3\}$

22. function; domain: $\{5, 3, 1, -1\}$; range: $\{2, -1, -3, -5\}$

23. function; domain: $\{4, 3, 2, 1, 0\}$; range: $\{2\}$

24. function; $(-\infty, \infty)$

25. not a function; $(-\infty, \infty)$

26. function; $[4, \infty)$

27. not a function; $[-1, \infty)$

28. function; $(-\infty, 0) \cup (0, \infty)$

29. function; $(-\infty, \infty)$

Objective 3

30. function

31. function

32. not a function

33. not a function

34. function

35. function

36. function

37. not a function

Objective 4

38. $1; 13; -2x + 5$

39. $10; -2; 6 + 2x$

40. $12; 48; 3x^2$

41. $8; 8; x^2 + 2x$

42. $\dfrac{4}{5}; \dfrac{4}{17}; \dfrac{4}{x^2 + 1}$

43. $-\dfrac{3}{5}; \dfrac{9}{5}; \dfrac{-2x + 1}{5}$

Objective 5

44. $f(x) = x + 2$

45. $f(x) = 2x + 2$

46. $f(x) = x - 1$

47. $f(x) = -\dfrac{1}{2}x - 2$

48. $f(x) = -\dfrac{1}{4}x + \dfrac{1}{2}$

49. $f(x) = 3x$

50. $f(x) = -x + 1$

51. $f(x) = -\dfrac{3}{2}x - 3$

3.6 Mixed Exercises

52. function; domain: $\{0, 1, 2, 4\}$; range: $\{1, 3, -4, -8\}$

53. function; domain: $\{-2, -3, 1\}$; range: $\{-5, -2\}$

54. not a function; domain: $\{-4, -5, -3, -2\}$; range: $\{1, 2, 3, 4, 5\}$

55. function; domain: $\{1, 2, 3, 4, 5\}$; range: $\{10, 9, 8, 7, -4\}$

56. function; domain: $(-\infty, \infty)$

57. function; domain $[-2, \infty)$

58. function; domain: $(-\infty, \infty)$

59. not a function; domain: $[-4, \infty)$

60. $f(x) = -\dfrac{x}{3} + 3; \dfrac{10}{3}; 1; -\dfrac{x}{3} + \dfrac{10}{3}$

61. $f(x) = 2x - 12; -14; 0; 2x - 14$

Chapter 4

SYSTEMS OF EQUATIONS AND INEQUALITES

4.1 Solving Systems of Linear Equations by Graphing

Objective 1

1. No	**2.** No	**3.** Yes
4. No	**5.** Yes	**6.** No
7. No	**8.** Yes	**9.** No
10. Yes	**11.** Yes	**12.** Yes
13. Yes	**14.** Yes	

Objective 2

15. $(2,-2)$	**16.** $(4,-1)$	**17.** $(-1,-2)$	**18.** $(-2,1)$
19. $(0,0)$	**20.** $(-9,2)$	**21.** $(4,8)$	**22.** $(-3,4)$
23. $(2,-1)$	**24.** $(4,-2)$	**25.** $(1,4)$	**26.** $(-2,6)$

Objective 3

27. No solution	**28.** Infinite number of solutions
29. No solution	**30.** Infinite number of solutions
31. Infinite number of solutions	**32.** No solution

4.1 Mixed Exercises

33. Yes	**34.** Yes	**35.** Yes	**36.** Yes
37. No	**38.** No	**39.** Infinite number of solutions	**40.** Infinite number of solutions
41. $(4,-2)$	**42.** $(2,1)$	**43.** $(-6,4)$	**44.** $(1,1)$

4.2 Adding Whole Numbers

Objective 1

1. $(1,6)$

2. $(2,4)$

3. $(7,-2)$

4. $(1,-5)$

5. $(2,7)$

6. $(-3,-5)$

7. $(-2,-1)$

8. $(4,-9)$

9. $(1,2)$

10. $(4,3)$

11. $(-2,2)$

12. $(5,1)$

13. $(4,-3)$

14. $\left(3,-\frac{2}{3}\right)$

15. Infinite number of solutions

16. $(1,-2)$

17. $(-2,-3)$

18. $\left(\frac{1}{2},-\frac{1}{2}\right)$

19. $(3,4)$

20. $(5,-2)$

21. $(4,-5)$

22. $\left(\frac{1}{2},3\right)$

23. $(0,0)$

24. $(4,5)$

25. $(1,2)$

26. $(4,-2)$

27. $(2,-1)$

28. $(-3,5)$

Objectives 2 and 3

29. No solution

30. Infinite number of solutions

31. No solution

32. $(6,2)$

33. $(-9,-11)$

34. $(2,3)$

35. $\left(\frac{22}{9},\frac{8}{9}\right)$

36. $(6,2)$

37. $(-12,0)$

38. $(0,4)$

39. Infinite number of solutions

40. $(-3,2)$

41. Infinite number of solutions

42. $\left(\frac{3}{5},\frac{1}{5}\right)$

43. $(2,-4)$

44. No solution

4.2 Mixed Exercises

45. $(-4,2)$

46. $(4,1)$

47. $(-4,-1)$

48. $(-4,2)$

49. Infinite number of solutions

50. $\left(\frac{1}{3},\frac{2}{5}\right)$

51. $(2,-4)$

52. $(-1,3)$

53. $(-4,4)$

54. $(0,-4)$

55. $(-2,2)$

56. $(-1,0)$

57. $(1,-1)$ **58.** $\left(\frac{3}{5},-\frac{2}{3}\right)$ **59.** Infinite number   **60.** $(9,-6)$
of solutions

61. $(12,-6)$ **62.** No solution

4.3 Solving Systems of Linear Equations by Elimination

Objective 1

1. $(1,4)$ 2. $(1,-2)$ 3. $(8,3)$ 4. $(4,-2)$

5. $(2,-4)$ 6. $(3,1)$ 7. $(0,7)$ 8. $(-1,-2)$

9. $(5,0)$ 10. $(-3,3)$ 11. $(4,-1)$ 12. $(-7,1)$

13. $\left(2,\frac{1}{6}\right)$ 14. $(2,-4)$ 15. $\left(\frac{1}{2},-\frac{1}{6}\right)$

Objective 2

16. $(4,-2)$ 17. $\left(\frac{1}{2},1\right)$ 18. $(3,-2)$ 19. $(-4,4)$

20. $(2,-2)$ 21. $(1,-5)$ 22. $(5,-4)$ 23. $(-3,2)$

24. $(4,1)$ 25. $(4,-5)$ 26. $\left(\frac{1}{2},-\frac{3}{2}\right)$ 27. $(2,3)$

28. $(1,-1)$ 29. $(0,-4)$ 30. $(2,5)$

Objective 3

31. $(4,-1)$ 32. $(-3,-2)$ 33. $(2,-4)$ 34. $(-4,1)$

35. $(5,-2)$ 36. $(3,-2)$ 37. $(4,-1)$ 38. $(-2,2)$

39. $(3,5)$ 40. $(5,15)$ 41. $(1,2)$ 42. $(-2,3)$

Objective 4

43. No solution   44. Infinite number of solutions

45. Infinite number of solutions   46. Infinite number of solutions

47. No solution   48. No solution

49. Infinite number of solutions   50. Infinite number of solutions

51. No solution   52. Infinite number of solutions

53. No solution   54. No solution

4.3 Mixed Exercises

55. $(4,2)$ **56.** $(-1,7)$ **57.** $(-4,3)$ **58.** $(-1,1)$

59. Infinite number of solutions **60.** No solution **61.** $(-4,-3)$ **62.** $(-4,3)$

63. $\left(-\frac{1}{3},2\right)$ **64.** $\left(\frac{2}{3},-\frac{11}{3}\right)$ **65.** Infinite number of solutions **66.** $\left(-\frac{5}{4},\frac{3}{4}\right)$

4.4 Applications of Linear Systems

Objective 1

1. 41 and 23
2. −91 and 25
3. 23 and 50
4. 8 and 12
5. 18 and 32
6. 5647, 3398
7. Carla, 28; Linda, 21
8. 26 cm; 56 cm
9. Length, 14 ft; width, 11 ft
10. 32 cm, 32 cm, 52 cm

Objective 2

11. 5000 adults; 3000 children
12. 300 regular; 100 student
13. 6 fives; 5 twenties
14. 1000 25-cent stamps; 250 15-cent stamps
15. 300 nonstudent; 111 student
16. 30 fives; 60 tens
17. 396 children's tickets; 327 adult tickets
18. 1500 general admission; 750 reserved
19. $6000 at 7%; $4000 at 4%
20. $4000 at 7%; $8000 at 9%

Objective 3

21. 30 pounds at $9 per pound;
    15 pounds at $6 per pound
22. 20 pounds at $1.60 per pound;
    10 pounds at $2.50 per pound
23. 40 bags at $90 a bag;
    10 bags at $75 a bag
24. 5 pounds of walnuts;
    5 pounds of peanuts
25. .6 pounds of caramels;
    .4 pounds of creams
26. 28 liters of 75% solution;
    42 liters of 55% solution
27. 10 liters of 90% solution;
    20 liters of 75% solution
28. 30 liters of 15% solution;
    15 liters of 30% solution
29. 70 liters of 10% solution;
    30 liters of 50% solution
30. 5 liters of 20% solution;
    10 liters of 5% solution

Objective 4

31. Rick, 48 mph; Hilary, 78 mph
32. 128 mph; 148 mph
33. Steve, 6 mph; Vic, 4 mph
34. 8 mph; 12 mph
35. Faustino, 80 mph; Pablo, 40 mph
36. John, 54 mph; Mike 52 mph

37. Canoe, 5 mph,; current, 3 mph

38. Wind, 35 mph; plane, 265 mph

39. Current, 5 mph; kayak, 11 mph

40. Plane A, 400 mph; Plane B, 360 mph

4.4 Mixed Exercises

41. Length, 43 cm; width, 29 cm

42. 8 at $14 each; 2 at $25 each

43. 9 fives; 5 tens

44. Bill, 642 km/hr; Monica, 582 km/hr

45. 75 milliliters

46. $15,000 at 6%; $35,000 at 7%

47. Current, 4 mph; boat, 12 mph

48. 50 pounds at $6 per pound;
100 pounds at $3 per pound

49. Enid, 76 mph; Hyman, 48 mph

50. 6 liters

4.5 Solving Systems of Linear Inequalities

Objective 1

1.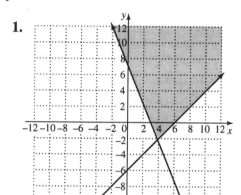

2.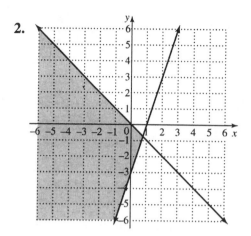

3.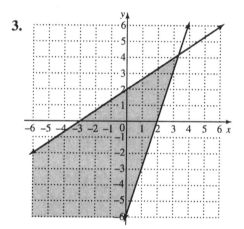

4.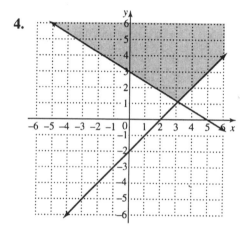

5.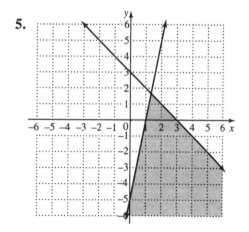

6.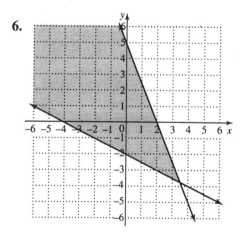

7.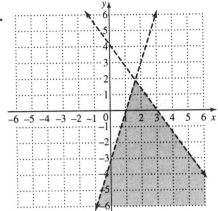

8.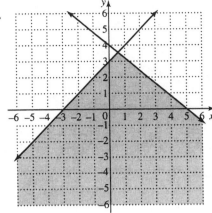

9.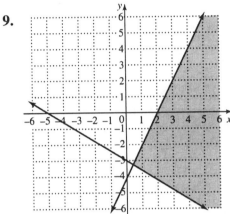

10.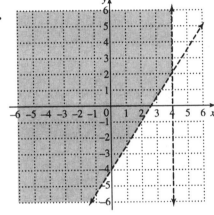

11.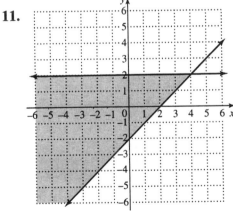

12.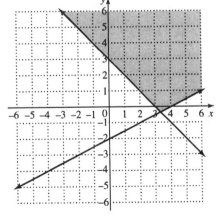

13.

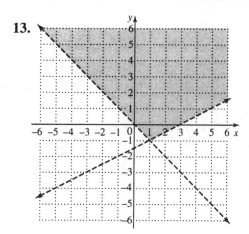

14.

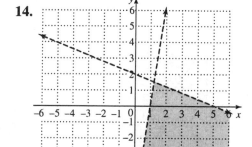

15.

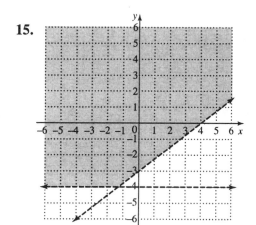

16.

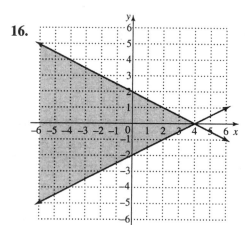

17.

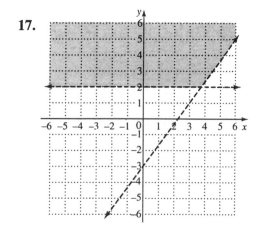

18.

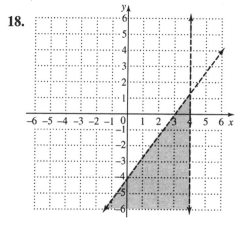

19.

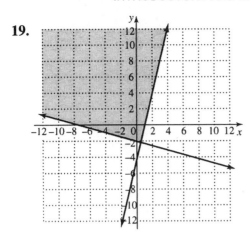

20.

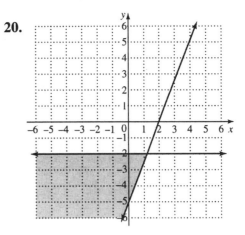

Chapter 5

EXPONENTS AND POLYNOMIALS

5.1   Addition and Subtraction of Polynomials

Objective 1

1. $6s^3$

2. $-7t^6$

3. $-3z^3$

4. $-7x^3 + 2x^2 + 6x + 4$

5. $-.4m^5$

6. $-.7r^2 - 8.2r$

7. $8c^3 - 8c^2 - 6c + 6$

8. $13y^4 - 7y^3 - 9y^2 - 3y + 4$

9. $-4q^3 - 6q^2$

10. $-19x^3 - 2x^2 + 4x$

11. $-\frac{1}{4}r^3$

12. $\frac{9}{10}m^3 - m^2$

Objective 2

13. (a) and (b)

14. (a)

15. (c)

16. (c)

17. (c)

18. (a) and (b)

19. $-m^2 - 4m + 6$; degree 2; trinomial

20. $m^4 + 7m^2 + 3m$; degree 4; trinomial

21. $3z^4 + 3z^3$; degree 4; binomial

22. $-6p^5 + 2p^3$; degree 5; binomial

23. $n^8 - n^2$; degree 8; binomial

24. $-3r^4 + 3r^2$; degree 4; binomial

25. $\frac{1}{2}x^2 - \frac{1}{2}x$; degree 2; binomial

26. $\frac{1}{2}y^2 - \frac{1}{2}y + \frac{5}{6}$; degree 2; trinomial

Objective 4

27. $2m^4 + 7m^3 - 7$

28. $5y^3 - 6y^2 + 13$

29. $5m^3 - 2m^2 - 4m + 4$

30. $7w^4 + m + 14$

31. $-x^5 - 2x^2 + 2x + 2$

32. $9y^3 - 9y^2 + 11$

33. $9p^4 - 3p^3 - 3p^2 - p + 22$

34. $-9z^4 - 3z^2 + 2z + 4$

35. $-3x^4 + 8x^3 - 3x^2 - 3$

36. $16x^3 - 2x^2 - 2x + 1$

37. $7y^4 + 4y^3 - y - 3$

38. $-4a^5 - 2a^3 - 2a - 4$

39. $4x^5 - 7x^4 + 3x^2 - x$

40. $m^4 - 2m^3 + 9m^2 - 4$

Objective 5

41. $6x^2 - 2x$

42. $-m^3 - 2m$

43. $4m^2 - 2$

44. $16k^2 + 2k - 22$

45. $-n^3 + 6n + 3$

46. $-x^5 - 6x^3 + 5x$

47. $19y^4 - 8y^3 + 4y^2 - 2y$

48. $2z^5 - z^4 - 2z^3 + 7z^2 - 8z + 7$

49. $3x^3 + 7x^2 + 12x + 2$

50. $-4x^3 - 4x^2 + 7x + 8$

51. $3p^2 + 10p - 6$

52. $4z^4 - 3z^3 + 8z + 7$

53. $5a^5 - 2a^3 + 3a^2 + 13$

54. $3p^4 - 9p^3 + 2p^2 + 7p - 7$

Objective 6

55. $2x^2y + 10y^2$; degree 3

56. $-3a^6 + a^4b - 3b^2$; degree 6

57. $-8m^2n - 4n - 8m$; degree 3

58. $-5ab - 6ac$; degree 2

59. $-7x^2y + 3xy + 5xy^2$; degree 3

60. $.11a^2 - .07b^2$; degree 2

61. $2x^3y - 2xy^3$; degree 4

62. $-2c^3d + 7c^2d - 9d^2$; degree 4

63. $-3a^3b + 6a^2b^2 + 9ab^3 + 11$; degree 4

64. $-11rs + 9rt - 4st$; degree 2

5.1 Mixed Exercises

65. $9b^3 + 4b^2 + 13b + 5$;
degree 3; none of these

66. $-2y^2 + 4y - 12$;
degree 2; trinomial

67. $2y^4 + 5y^3 + 11y + 2$;
degree 4; none of these

68. $6m^3 + 5m^2 - 3m + 10$;
degree 3; none of these

69. $5x^5 - 3x^4 - 2x$;
degree 5; trinomial

70. $4p^4 - 5p^3 + 8p - 4$;
degree 4; none of these

71. $11y^2 + 2y + 6$;
degree 2; trinomial

72. 1; degree 0; monomial

73. $4x^3 - 5$; degree 3; binomial

74. $4m^2 + m - 12$; degree 2
trinomial

5.2   The Product Rule and Power Rules for Exponents

Objective 1

1. $2^6 = 64$

2. $10^3 = 1000$

3. $(-1)^8 = 1$

4. $(.4)^3 = .064$

5. $\left(\frac{1}{3}\right)^5 = \frac{1}{243}$

6. $\left(-\frac{2}{5}\right)^4 = \frac{16}{625}$

7. r^5

8. $(-2y)^5$

9. $\left(\frac{1}{4}mn\right)^4$

10. $(.5st)^4$

11. 256; base, −4; exponent, 4

12. −81; base, 3; exponent, 4

13. 64; base 2; exponent, 6

14. 625; base, −5, exponent, 4

15. −36; base 6 exponent, 2

16. $\frac{1}{81}$; base, $-\frac{1}{3}$, exponent, 4

Objectives 2

17. 4^{11}

18. $(-3)^{13}$

19. $(.1)^{14}$

20. Cannot use product rule

21. $\left(\frac{1}{2}\right)^{10}$

22. $\left(\frac{2}{5}\right)^{10}$

23. $(-5)^{14}$

24. Cannot use product rule

25. $54p^9$

26. $100n^{12}$

27. $84a^8$

28. $-144b^{12}$

29. $8r^6$

30. $-32x^{15}$

31. $16y^2$; $63y^4$

32. $-7m^6$; $12m^{12}$

33. $-2a^3$; $20a^9$

34. $8t^2$; $-48t^6$

Objectives 3

35. 3^{12}

36. 7^{27}

37. 6^{24}

38. $\left(\frac{1}{3}\right)^{15}$

39. 8^{35}

40. 9^{45}

41. $(-3)^{21}$

42. $(-2)^{35}$

43. -13^{24}

44. 14^{25}

45. 6^{110}

46. -21^{20}

Objectives 4

47. $\frac{1}{9}x^8$

48. $y^3 z^{12}$

49. $p^8 q^{12}$

50. $r^{14} s^4$

51. $5a^3 b^9$

52. $2^4 w^{12} z^{28}$

53. $5^4 r^{12} t^8$

54. $-.008 a^{12} b^3$

55. $(-2)^6 r^{18} s^6$

56. $2 \cdot 3^4 c^{16} d^{28}$

57. $-162 x^4$

58. $4^3 c^9 d^{12}$ or $64 c^9 d^{12}$

Objectives 5

59. $\dfrac{3^3}{5^3}$

60. $\dfrac{5^2}{8^2}$

61. $\dfrac{(-2)^7 a^7}{b^{14}}$

62. $\dfrac{w^4}{2^4}$

63. $\dfrac{z^6}{3^3}$

64. $\dfrac{(-2)^2 x^2}{5^2}$

65. $\dfrac{x^4 y^4}{z^8}$

66. $\dfrac{x^{27} y^9}{z^{36}}$

67. $\dfrac{(-2)^3 x^3}{5^3}$

68. $\dfrac{2^5 a^{10}}{b^{15}}$

69. $\dfrac{x^4}{16 y^4}$

70. $\dfrac{4^3}{g^3}$

Objectives 6

71. $7^9 a^9$

72. $(-4)^9 q^9$

73. $\dfrac{4^{13}}{7^9}$

74. $\dfrac{3^4 b^8}{11^4}$

75. $\dfrac{3^3 \cdot 5^4 x^4}{4^3}$

76. $\dfrac{4^3 a^3 b^6}{3^3}$

77. x^{26}

78. $5^{11} x^{18} y^{37}$

79. $\dfrac{7^7 a^{14} b^{21}}{2^7}$

80. $\dfrac{k^7 m^{28} p^{14}}{3^7 n^{28}}$

81. $(-2)^9 p^9 q^9$

82. $32 a^9 b^{14} c^5$

5.2 Mixed Exercises

83. $\frac{1}{81}$; base, $-\frac{1}{3}$; exponent 4

84. -64; base, 2; exponent, 6

85. -1; base, -1; exponent, 5

86. -1; base, 1; exponent, 7

87. -125; base, 5; exponent 3   **88.** 16; base, -2; exponent 4

89. 6^{12}

90. $x^{28}y^{7}$

91. $3^{4}a^{28}b^{12}$

92. $\dfrac{3^{8}x^{8}}{5^{8}y^{8}}$

93. $3^{4}m^{12}n^{16}$

94. $4^{12}w^{12}t^{12}$

95. $(-4)^{5}a^{16}$

96. $7^{3}a^{10}b^{18}c^{19}$

97. z^{27}

98. $\dfrac{3^{6}a^{8}}{4^{2}}$

99. $\dfrac{7^{4}w^{4}x^{20}}{y^{8}}$

100. $\dfrac{r^{12}s^{4}t^{4}}{2^{4}n^{8}}$

5.3 Multiplying Polynomials

Objective 1

1. $12y^5$

2. $16y^7$

3. $35z^3 + 14z$

4. $-6x^4 - 12x^5 - 4x^6$

5. $-6z^6 - 18z^4 - 24z^2 - 12z$

6. $-6y^5 - 9y^4 + 12y^3 - 33y^2$

7. $12k^2 + 8k^5 + 24k^6$

8. $6m + 14m^3 + 6m^4$

9. $-35b^4 + 7b^2 - 28b^3$

10. $-8r^6 + 12r^5 - 8r^4$

11. $32m^3n + 16m^2n^2 + 56mn^3$

12. $-24r^4s^5 + 12r^3s^4 - 6r^3s^5$

Objective 2

13. $x^2 + 12x + 27$

14. $y^2 - 6y + 8$

15. $3p^2 + 10p + 8$

16. $28 + 29a + 6a^2$

17. $10n^2 + 27n + 5$

18. $6x^2 + 23x + 20$

19. $6m^2 + 2m - 20$

20. $x^3 + 27$

21. $y^3 + 64$

22. $2r^3 + 3r^2 - 4r + 15$

23. $9y^4 - 18y^3 + 5y^2 + 16y - 16$

24. $6m^5 + 4m^4 - 5m^3 + 2m^2 - 4m$

25. $2z^6 + 4z^5 - z^4 + z^2 - 9z - 6$

26. $4x^4 + 5x^2 + 3x + 12$

27. $4x^3 - 5x + 6$

28. $6y^4 - 5y^3 - 8y^2 + 7y - 6$

29. $6a^5 - 4a^4 + 5a^3 - 2a^2 + a$

30. $4b^6 - 8b^5 + b^4 + 2b^3 - 11b^2 + 3b + 6$

31. $8x^5 + 12x^4 + 12x^3 + 12x^2 + 13x + 6$

32. $-9y^6 + 3y^5 - 9y^4 + 3y^3 - 2y^2 - 6y + 4$

33. $a^2 + 10a + 25$

34. $4r^2 + 4rs + s^2$

35. $-4m^6 - 14m^4 + 8m^3 - 12m^2 + 15m - 3$

36. $6x^4 + 11x^3 - 9x^2 - 4x$

37. $4x^4 - 4x^3 + 9x^2 - 4x + 4$

38. $y^4 - 4y^3 - 2y^2 + 12y + 9$

Objective 3

39. $4m^2 - 25m - 21$

40. $6x^2 - 5xy - 6y^2$

41. $20a^2 + 11ab - 3b^2$

42. $121k^2 - 16$

43. $3 + 10a + 8a^2$

44. $4x^2 + 5xy - 6y^2$

45. $-6m^2 - mn + 12n^2$

46. $2v^4 - 5v^2w^2 - 3w^4$

47. $x^2 - .2x - .15$

48. $4y^2 - .8y - .05$

49. $x^2 - \frac{5}{3}x + \frac{4}{9}$

50. $z^2 + \frac{2}{5}z - \frac{8}{25}$

5.3 Mixed Exercises

51. $28x^7$

52. $-10r^7 + 20r^6 - 15r^4$

53. $a^2 + 2a - 8$

54. $27p^2 - 3pr - 10r^2$

55. $15t^2 + 4tu - 4u^2$

56. $18y^5 - 6y^4 - 9y^3 + 12y^2$

57. $x^2 - 16y^2$

58. $42p^4 - 24p^5 + 12p^7 - 24p^9$

59. $3x^4 + 4x^3 - 4x^2 + 4x + 8$

60. $3x^3 - 14x^2 + 5x + 12$

61. $x^4 - 8x^3 + 22x^2 - 24x + 9$

62. $2x^5 - x^4 - 8x^3 + 10x^2 - 10x + 3$

63. $6x^5 + 3x^4 - x^3 + 10x^2 - 5x + 2$

5.4 Special Products

Objective 1

1. $z^2 + 6z + 9$

2. $t^2 + 8t + 16$

3. $a^2 + 4ab + 4b^2$

4. $25y^2 - 30y + 9$

5. $4m^2 + 20m + 25$

6. $25m^2 + 30mn + 9n^2$

7. $49 + 14x + x^2$

8. $25 + 20y + 4y^2$

9. $4p^2 + 12pq + 9q^2$

10. $4m^2 - 12mp + 9p^2$

11. $16y^2 - 5.6y + .49$

12. $16x^2 - 2xy + \frac{1}{16}y^2$

13. $9x^2 - 2xy + \frac{1}{9}y^2$

14. $9a^2 + 3ab + \frac{1}{4}b^2$

Objective 2

15. $z^2 - 36$

16. $144 - x^2$

17. $49x^2 - 9y^2$

18. $64k^2 - 25p^2$

19. $16p^2 - 49q^2$

20. $4 - 9x^2$

21. $81 - 16y^2$

22. $x^2 - .04$

23. $y^2 - \frac{16}{9}$

24. $49m^2 - \frac{9}{16}$

25. $4a^2 - \frac{16}{9}b^2$

26. $\frac{9}{16}s^2 - \frac{49}{25}t^2$

27. $y^4 - 4$

28. $25m^8 - 49n^6$

Objective 3

29. $x^3 + 6x^2 + 12x + 8$

30. $a^3 - 9a^2 + 27a - 27$

31. $y^3 + 12y^2 + 48y + 64$

32. $8x^3 - 36x^2 + 54x - 27$

33. $8x^3 + 12x^2 + 6x + 1$

34. $k^4 + 8k^3 + 24k^2 + 32k + 16$

35. $t^4 - 12t^3 + 54t^2 - 108t + 81$

36. $x^4 + 8x^3y + 24x^2y^2 + 32xy^3 + 16y^4$

37. $81b^4 - 216b^3 + 216b^2 - 96b + 16$

38. $256s^4 + 768s^3t + 864s^2t^2 + 432st^3 + 81t^4$

39. $8x^3 - 12x^2 + 6x - 1$

40. $j^4 + 12j^3 + 54j^2 + 108j + 81$

5.4 Mixed Exercises

41. $16y^2 + 40y + 25$

42. $n^2 - 2n - 15$

43. $16b^2 - 81$

44. $9b^2 - 30b + 25$

45. $49t^2 - 36u^2$

46. $3p^2 + 10p - 8$

47. $21h^2 + 5hk - 6k^2$

48. $25k^2 + 30km + 9m^2$

49. $25 - 36w^2$

50. $32a^2 + 28a - 15$

51. $16x^2 - \frac{49}{16}$

52. $4y^2 - 1.2y + .09$

53. $16j^2 + 4jk + \frac{1}{4}k^2$

54. $\frac{16}{49}t^2 - 4u^2$

5.5   Integer Exponents and the Quotient Rule

Objective 1

1. 1   **2.** −1   **3.** −1   **4.** 1   **5.** 2   **6.** 2

7. 1   **8.** 0   **9.** 0   **10.** −2   **11.** −1   **12.** 0

Objective 2

13. $\frac{1}{16}$ **14.** $\frac{7}{2}$ **15.** $-\frac{1}{27}$ **16.** $\frac{1}{16}$ **17.** $\frac{25}{9}$

18. $\frac{9}{2}$ **19.** $\frac{3}{8}$ **20.** $\frac{1}{20}$ **21.** $\frac{1}{r^7}$ **22.** $\frac{1}{a^4}$

23. $2r^7$ **24.** $\frac{1}{6r^2}$ **25.** $\frac{2y^7}{3x^4}$ **26.** 4

Objective 3

27. 16   **28.** 9^5 **29.** $(-2)^5$ **30.** $\frac{1}{(-4)^6}$

31. $\frac{1}{2x^6}$ **32.** $\frac{k^4 m^5}{2}$ **33.** $\frac{x^6}{12^3 y^2}$ **34.** 27

35. $\frac{1}{12}$ **36.** $a^6 b^6$ **37.** $\frac{p^8}{3^5 m^3}$ **38.** $8^5 b^4 c^7$

Objective 4

39. 7^6 **40.** 9   **41.** 6^4 **42.** x^9

43. $\frac{1}{2^{16}}$ **44.** $\frac{3^{16} y^4}{x^8}$ **45.** $\frac{4^2}{5^2 w^2 y^{10}}$ **46.** $\frac{q^5}{p^{10}}$

47. $\frac{3^2}{2^4 y^2}$ **48.** $\frac{1}{9xy}$ **49.** c^{25} **50.** $a^{16} b^{22}$

51. $\frac{1}{x^4 y^{12}}$ **52.** $\frac{1}{k^4 t^{20}}$ **53.** $\frac{2^4 3^8 x^{20}}{y^8}$ **54.** $\frac{1}{4^3 a^5 b^9}$

5.5 Mixed Exercises

55. 2

56. $4\frac{1}{4}$

57. 0

58. 1

59. 1

60. $\frac{27}{8}$

61. 7^2

62. $\dfrac{1}{a^{12}}$

63. $b^6 c^5$

64. 8^3

65. $\frac{64}{27}$

66. $\dfrac{16}{k^6}$

67. $\dfrac{x^6}{y^5}$

68. $\dfrac{1}{x^3}$

69. $\dfrac{2}{3x^6}$

70. $\dfrac{s^2}{r^6}$

71. y^5

72. $\dfrac{4n^2 p^4}{m}$

5.6   Dividing a Polynomial by a Monomial

Objective 1

1. $4m+2$

2. $8m^2-2m+3$

3. $7m^3-1$

4. $9m^4+6m^2-11-\dfrac{2}{m^2}$

5. $7m+12-\dfrac{4}{m}$

6. $8m-\frac{1}{2}$

7. $-21m+9+\dfrac{2}{m}$

8. $1-\dfrac{1}{m^2}$

9. $8m-\dfrac{12}{m}-\dfrac{3}{m^2}$

10. $4m^3-5m^2+1-\dfrac{1}{2m}+\dfrac{3}{4m^2}$

11. $1+3p^3$

12. $3x^4+7x^3+5x$

13. $5a^2-\frac{9}{4}$

14. $2z^4+9z^2-4+\dfrac{10}{3z}$

15. $10x^3-5x$

16. $3x^3+8x^2-16+\dfrac{4}{x}$

17. $\dfrac{1}{2}m+\dfrac{7}{2}-\dfrac{21}{m}$

18. $8p^2-7p-\dfrac{3}{p}$

19. $7q^2-4+\dfrac{1}{q}$

20. $2y^6+8y^3-41-\dfrac{12}{y^3}$

21. $3z^2+7z-2+\dfrac{3}{4z^2}$

22. $12+16x^3+\frac{1}{2}x^7$

23. $-5u^2+4uv-9v^2$

24. $-3+\dfrac{2}{y}-\dfrac{6}{y^2}$

5.7 Dividing a Polynomial by a Polynomial

Objective 1

1. $x+2$

2. $y-6$

3. $6a-5$

4. $p+8$

5. $x+8$

6. $r+3+\dfrac{-5}{r-5}$

7. $a-7+\dfrac{37}{2a+3}$

8. $3w+2$

9. $5w-2+\dfrac{-4}{w-4}$

10. $5b-3+\dfrac{24}{b+7}$

11. $3m-2+\dfrac{8}{3m-4}$

12. $9a-1$

13. $2x+5$

14. $3y^2-5y+6$

15. $z^2-5z+9+\dfrac{-25}{2z+3}$

16. $2m^2+m-5+\dfrac{26}{3m+2}$

17. $9p^3-2p+6$

18. $3x-2+\dfrac{34x-35}{x^2-3x-5}$

19. $3x^2-6x+2+\dfrac{13x-7}{2x^2+3}$

20. $3y^3-2y^2+2y-1+\dfrac{6y-8}{4y^2-3}$

21. a^2+1

22. $3x^2+2x+1+\dfrac{-1}{x^2-1}$

23. y^2-y+1

24. b^2+1

25. $2x^4+x^3-3x^3+2x+1+\dfrac{2}{3x+2}$

26. $16x^4+24x^3+36x^2+54x+81$

5.8 An Application of Exponents: Scientific Notation

Objective 1

1. 3.25×10^2

2. 4.579×10^3

3. 2.3651×10^4

4. 2.09907×10^5

5. 7.42×10^0

6. 4.296×10^{11}

7. 2.57×10^{-2}

8. 2.46×10^{-1}

9. 4.13×10^{-6}

10. 4.26×10^{-3}

11. -4.3276×10^4

12. -4.7×10^{-4}

Objective 2

13. 25,000

14. 72,000,000

15. $-2,450,000$

16. 4.045

17. 23,000

18. 45,000,000

19. .0064

20. .000724

21. .04007

22. .4752

23. -4.02

24. $-.000911$

Objective 3

25. 210,000,000

26. 24,000

27. 253

28. .00000713

29. .02

30. 50,000

31. 4,000,000

32. .0313

33. 200

34. 6000

35. 2.1

36. 9

5.8 Mixed Exercises

37. 2.705×10^6

38. 4.6×10

39. 2.53×10^{-2}

40. 7.5×10^{-1}

41. -4.327×10^0

42. 4.03×10^{-5}

43. 544

44. .014

45. 15,000

46. 10

47. 210,000

48. 3000

49. 200

50. .004

51. 350

52. 1600

53. 100

54. 10

Chapter 6

FACTORING AND APPLICATIONS

6.1 Factors; The Greatest Common Factor

Objective 1

1. 3 **2.** 6 **3.** 5 **4.** 12 **5.** 15

6. 1 **7.** 3 **8.** 14 **9.** 32 **10.** 28

Objective 2

11. $5x^2$ **12.** $9b^3$ **13.** $2ab^2$ **14.** m^4 **15.** $6a^6$

16. z^2 **17.** $k^2m^4n^4$ **18.** $w^2x^5y^4$ **19.** $15a^2y$ **20.** $9xy^2$

Objective 3

21. 21 **22.** $6x$ **23.** $6y^3$

24. $-3a$ **25.** $6(3r+4t)$ **26.** $13(2r+3t)$

27. $9q(2q-3)$ **28.** $9x(5y+2+3x^2y)$ **29.** $10x(2x+4xy-7y^2)$

30. $8a(3b-a+5c)$ **31.** $21t(2w+1-3t)$ **32.** $11r(4s-2+7r)$

33. $5a^3(3a^4-5-8a)$ **34.** No common factor (except 1) **35.** $13x^8(2-x^4+4x^2)$

36. $27r^2(1-2r^2-3r^3)$ **37.** $8xy^2(7xy^2-3y+4)$ **38.** $(a+b)(3-x)$

39. $(9+a)(x+8)$ **40.** $(r-4s)(x^2+z^2)$

Objective 4

41. $(x+2)(x+5)$ **42.** $(x^2+2)(x^2+5)$ **43.** $(x-4)(x+5)$

44. $(x+9)(x-4)$ **45.** $3(x-3)(x+4)$ **46.** $(2x+3y)(4x-y)$

47. $(x-2)(y-2)$ **48.** $(3-x)(5-y)$ **49.** $(x+3)(5-y)$

50. $\left(2x+3y^2\right)\left(3x^2-y^3\right)$ **51.** $\left(2a-3b\right)\left(a^2+b^2\right)$ **52.** $\left(3x^2-y\right)\left(4x-y^2\right)$

53. $\left(2x^2+3y\right)\left(x^2+2y^2\right)$ **54.** $2\left(3x+y\right)\left(2x-y\right)$

6.1 Mixed Exercises

55. 7

56. 6

57. $5ab^2$

58. $17pq^2$

59. No common factor (except 1)

60. $-9uv^2$

61. $6x\left(y^2-7z^2\right)$

62. $4q^2\left(pq^2+9\right)$

63. $15ab\left(3ab^2+b-6\right)$

64. $\left(x-2y\right)\left(2a+9b\right)$

65. $\left(m-5\right)\left(m+3\right)$

66. $\left(2x+y\right)\left(x-7y\right)$

67. $\left(5m+3\right)\left(3m-1\right)$

68. $\left(3r-2s\right)\left(r^2+s^2\right)$

69. $2p\left(3r^2+2q-4p\right)$

70. $\left(1+p\right)\left(1-q\right)$

6.2 Factoring Trinomials

Objective 1

1. 1 and 42, –1 and –42, 2 and 21, -2 and –21, 7 and 6, –7 and –6, 14 and 3, –14 and –3; the pair with sum of 17 is 3 and 14.

2. 1 and 28, –1 and –28, 2 and 14, –2 and –14, 4 and 7, –4 and –7; the pair with a sum of –11 is –4 and –7.

3. –8 and 8, 1 and –64, –1 and 64, 2 and –32, –2 and 32, 16 and –4, –16 and 4; the pair with a sum of 12 is –4 and 16.

4. 1 and –54, –1 and 54, 2 and –27, –2 and 27, 3 and –18, –3 and 18, 6 and –9, –6 and 9; the pair with a sum of –3 is 6 and –9.

5. $x+4$

6. $x+7$

7. $x+2$

8. $x-6$

9. $(x+2)(x+9)$

10. $(x-4)(x-7)$

11. $(x-2)(x+1)$

12. $(x+7)^2$ or $(x+7)(x+7)$

13. $(x-7)(x+5)$

14. $(x-11)(x+3)$

15. $(x+5)(x+1)$

16. $(x-7y)(x-8y)$

17. $(x-7y)(x+3y)$

18. $(m-3n)(m+n)$

Objective 2

19. $2(x-2)(x+7)$

20. $3(x+4)(x-2)$

21. $3hk(h+2)(h-9)$

22. $7(b-2)(b-4)$

23. Prime

24. $3p^4(p+2)(p+4)$

25. $2ab(a-3b)(a-2b)$

26. $3y(y-1)(y+4)$

27. $5(r+4)(r+3)$

28. $3x(y-2)(y-6)$

29. $10k^4(k+2)(k+5)$

30. $x^3(x-1)(x-2)$

31. $2y^2(x+2y)(x-3y)$

32. $b(a-7b)(a-5b)$

6.2 Mixed Exercises

33. $(a-7)(a-3)$

34. $(p-4)(p+3)$

35. $x(x-2)(x-2)$

36. $5r(r-7)(r-2)$

37. Prime

38. $2m(m-2)(m+1)$

39. $2n^2(n-5)(n-3)$

40. $(q-6)(q+2)$

41. $(b+5c)^2$ or
$(b+5c)(b+5c)$

42. $qr(r-7q)(r+3q)$

43. $(a+2b)(a+8b)$

44. $3(d-3)^2$ or
$3(d-3)(d-3)$

45. $2t(s+2)(s-5)$

46. $2x(x-5y)(x-2y)$

6.3 Factoring Trinomials by Grouping

Objective 1

1. $x+3$

2. $2x+5$

3. $4x-2$

4. $8y-3$

5. $(4b+3)(2b+3)$

6. $(x+2)(3x+7)$

7. $(5a+2)(3a+2)$

8. $(3n+4)(2n+1)$

9. $(3b+2)(b+2)$

10. $(3m+4)(m-3)$

11. $p(3p+2)(p+2)$

12. $2(4m+n)(m+3n)$

13. $b(7a+4)(a+2)$

14. $(2s-t)(s+3t)$

15. $3(c+2d)(3c+2d)$

16. $(5a+3b)(5a+3b)$

6.4 Factoring Trinomials Using FOIL

Objective 1

17. $(5x+2)(2x+3)$ 18. $(4y-5)(y+2)$ 19. $(a+6)(2a+1)$

20. $(w-2)(5w+1)$ 21. $(2q+1)(4q+3)$ 22. $(4m+1)(2m-3)$

23. $(2b+1)(7b-2)$ 24. $(5q+6)(3q-4)$ 25. $(3a+2b)(a+2b)$

26. $(3w+2z)^2$ or $(3w+2z)(3w+2z)$ 27. $(5c+2d)(2c-d)$ 28. $(2x+3y)(3x-4y)$

29. $(6x-y)(3x-4y)$ 30. $(4y-3)(3y+5)$

6.3 and 6.4 Mixed Exercises

31. $(3x+1)(x-4)$ 32. $(2p+1)(p+5)$ 33. $(3y-1)(2y+1)$

34. $(9y+2)(y-2)$ 35. $(p+5)(3p+2)$ 36. $(3r-1)(3r+5)$

37. $(x+4)(7x-1)$ 38. $2(2c-d)(c+4d)$ 39. $x^2(2x-3)(x+4)$

40. $2a(3a+2b)(2a+3b)$ 41. $(9r-4t)(3r+2t)$ 42. $y^3z^2(2y+z)(y-3z)$

6.5 Factoring Techniques

Objective 1

1. Prime

2. $(x+7)(x-7)$

3. $(10r+3s)(10r-3s)$

4. $(y+8)(y-8)$

5. $(5a+6)(5a-6)$

6. $\left(3j+\frac{4}{7}\right)\left(3j-\frac{4}{7}\right)$

7. $(6+11d)(6-11d)$

8. $(11m+3n)(11m-3n)$

9. $\left(x^2+9\right)(x+3)(x-3)$

10. $\left(3m^2+1\right)\left(3m^2-1\right)$

11. $\left(3x^2+4\right)\left(3x^2-4\right)$

12. $\left(9y^2+1\right)(3y+1)(3y-1)$

13. Prime

14. $m^2(mn+1)(mn-1)$

Objective 2

15. $(y+3)^2$

16. $(q+7)^2$

17. $(m-4)^2$

18. $(c+11)^2$

19. $\left(z-\frac{2}{3}\right)^2$

20. $(2w+3)^2$

21. $(4q-5)^2$

22. $(3j+2)^2$

23. $\left(8p^2+3q^2\right)^2$

24. $\left(10p-\frac{5}{8}r\right)^2$

25. $(3m+.1)^2$

26. $-4(2x+3)^2$

27. $-3(2a-5b)^2$

28. $2(3x+7y)^2$

Objective 3

29. $(x-y)(x^2+xy+y^2)$

30. $(2a-1)(4a^2+2a+1)$

31. $(2r-3s)(4r^2+6rs+9s^2)$

32. $(4x-y)(16x^2+4xy+y^2)$

33. $(6m-5p^2)(36m^2+30mp^2+25p^4)$

34. $(2a-5b)(4a^2+10ab+25b^2)$

35. $(r+s-1)(r^2+2rs+s^2+r+s+1)$

36. $2n(3m^2+n^2)$

37. $3x^2-3x+1$

38. $8(3x-y)(9x^2+3xy+y^2)$

Objective 4

39. $(x+y)(x^2-xy+y^2)$

40. $(z+2)(z^2-2z+4)$

41. $(3r+2s)(9r^2-6rs+4s^2)$

42. $8(a+2b)(a^2-2ab+4b^2)$

43. $(5p+q)(25p^2-5pq+q^2)$

44. $(4x+7y)(16x^2-28xy+49y^2)$

45. $(1+y+z)(1-y-z+y^2+2yz+z^2)$

46. $2x(x^2+3y^2)$

47. $(2a-1)(a^2-a+1)$

48. $2(t+1)(t^2+2t+4)$

6.5   Mixed Exercises

49. $(a+7)^2$

50. $(4a+5b)(4a-5b)$

51. $(3p+11)(3p-11)$

52. $(2f+5)^2$

53. Prime

54. $\left(7x-\frac{3}{2}\right)^2$

55. $(a^2-4b)^2$

56. $(z-.8)^2$

57. $(j+4)^2$

58. $\left(\frac{1}{3}x-6y\right)^2$

59. $(r^2+9x^2)(r+3x)(r-3x)$

60. $-3(x+4)^2$

61. $-5(2y+1)^2$

62. $3(4m^2+25)$

63. $(5m+2m)(25m^2-10mn+4n^2)$

64. $(y^2+1)(y^4-y^2+1)$

65. $(5m-4p)(25m^2+20mp+16p^2)$

66. $(z-5y)(z^2+5yz+25y^2)$

6.6 Solving Quadratic Equations by Factoring

Objective 1

1. $-9, \frac{3}{2}$

2. $-\frac{4}{3}, \frac{7}{5}$

3. $-2, -\frac{1}{3}$

4. $-7, 7$

5. $-\frac{5}{2}, 4$

6. $-7, 9$

7. $0, 3$

8. $3, -\frac{2}{3}$

9. $-1, \frac{3}{8}$

10. $-\frac{2}{3}$

11. $0, \frac{4}{5}$

12. $\frac{4}{3}, -\frac{4}{3}$

13. $-\frac{4}{3}, \frac{3}{4}$

14. $-\frac{2}{7}, \frac{3}{2}$

15. $-4, \frac{3}{5}$

16. $-3, \frac{1}{2}$

Objective 2

17. $0, -7, 2$

18. $0, -\frac{3}{2}, 5$

19. $-\frac{3}{2}, 0, \frac{3}{2}$

20. $-7, 0, 7$

21. $-5, 0, 5$

22. $-4, 0, 2$

23. $-2, 0, \frac{3}{2}$

24. $-\frac{3}{2}, \frac{3}{2}, 2$

25. $-9, 0, 1$

26. $-\frac{4}{3}, 0, 1$

27. $-4, -5, -2$

28. $-6, 2, 3, 6$

29. $0, 7, 8$

30. $7, -5, \frac{3}{2}$

31. $-\frac{5}{2}, \frac{3}{2}, -3$

32. $-5, 5, 1$

6.6 Mixed Exercises

33. $\frac{2}{3}, 7$

34. $-5, 0, 3$

35. $-5, -2$

36. $-9, 0$

37. $0, -2$

38. $-\frac{2}{3}, 6$

39. $-\frac{5}{3}, \frac{5}{3}$

40. $-7, 0, 7$

41. $-3, 3, 4$

42. $\frac{5}{3}, 7$

43. $0, 2, 4$

44. $0, -\frac{16}{3}$

45. $-\frac{2}{3}, 4$

46. $-2, 0, 2, 12$

6.7 Applications of Quadratic Equations

Objective 1

1. Length, 17 cm; width, 9 cm

2. Length, 21 inches; width, 7 inches

3. Length, 18 ft; width, 14 ft

4. First rectangle: length, 12 m; width, 4 m; Second rectangle: length, 8 m; width, 6 m

5. First square, 5 m; second square, 3 m

6. Base, 12 cm; height, 7 cm

7. Length, 9 cm; width, 3 cm

8. Length, 24 cm; width, 8 cm

9. 6 ft

10. 3 m

Objective 2

11. 0, 1 or 7, 8

12. 6, 7 or –6, –7

13. 7, 9

14. –2, –1

15. –4, –3, or 3, 4

16. 6, 8

17. 7, 9 or –7, –5

18. 12, 14

19. 6, 8

20. 6, 7 or –1, 0

Objective 3

21. 6 inches

22. 10 m, 24 m, 26 m

23. 7 m, 24 m, 25 m

24. Train, 80 mi; car, 60 mi

25. 15 ft

26. 30 ft

27. 18 ft

28. 45 m, 60 m, 75 m

29. 20 mi

30. 16 ft

6.7 Mixed Exercises

31. Length, 16 cm; width, 4 cm

32. Length, 12 ft; width, 6 ft

33. 8, 10

34. 11, 13, 15

35. 5 inches

36. Length, 8 ft; width, 5 ft

37. 4, 5

38. 45 mi

39. Carla, 24 miles; Penny, 10 miles

40. Length, 17 cm; width, 5 cm

41. Width, 8 yd

42. 26 mi

43. 5, 7, 9

44. 6 m and 8 m

45. (a)   1.5 sec
    (b)   1 sec and 2 sec

46. (a)   .5 sec
    (b)   1 sec
    (c)   2 sec

Chapter 7

RATIONAL EXPRESSIONS AND APPLICATIONS

7.1   Rational Expressions and Functions; Multiplying and Dividing

Objectives 1 and 2

1. 5

2. –5

3. –7

4. 0

5. None

6. $\dfrac{7}{4}$

7. $\dfrac{2}{3}$

8. 5, –5

9. 1, 2

10. 5

11. None

12. None

Objective 3

13. $\dfrac{n^2}{4}$

14. $\dfrac{7p^5}{3}$

15. $-\dfrac{x^2 y^3}{3}$

16. $\dfrac{m-7}{m-2}$

17. $\dfrac{1}{5}$

18. $\dfrac{3}{4}$

19. $\dfrac{6y+1}{3y+1}$

20. $\dfrac{x-3}{x+3}$

21. $\dfrac{11r^2}{6}$

22. $\dfrac{s+2}{s+4}$

23. $\dfrac{2z+3}{4z+3}$

24. 1

25. –1

26. $-(x+1)$ or $-x-1$

27. –1

28. $\dfrac{r^2 + rs + s^2}{r+s}$

29. $\dfrac{x+y}{x^2 + xy + y^2}$

30. $\dfrac{p-5}{5+p}$

Objective 4

31. $\dfrac{3x}{4}$

32. $\dfrac{3z}{2}$

33. $\dfrac{3}{8}$

34. $8y$

35. $2a^4$

36. $\dfrac{6}{m+3}$

37. $\dfrac{6}{5}$

38. 1

39. $\dfrac{3}{10}$

40. $\dfrac{2x}{3}$

41. $\dfrac{x+4}{x-4}$

42. $\dfrac{2z-3}{2z+3}$

Objective 5

43. $\dfrac{y}{3}$

44. $\dfrac{p-5}{4}$

45. $\dfrac{7x}{x^2 + 9}$

46. $\dfrac{5}{m^2 + 2m + 3}$

47. $\dfrac{p^2 + 7p}{2p-1}$

48. $\dfrac{7}{n+8}$

49. –1

50. $\dfrac{5+r}{r^2 + 2r}$

51. $\dfrac{3x-6}{x^2+4}$

52. $\dfrac{z^2-9}{7z+7}$

53. No reciprocal for 0

54. $\dfrac{x^2+x+2}{x^2-3x+4}$

Objective 6

55. $\dfrac{1}{4}$

56. $\dfrac{25t^3}{9}$

57. $\dfrac{8s^2}{3}$

58. $\dfrac{2r^3}{3}$

59. 2

60. $\dfrac{3}{2}$

61. $\dfrac{2}{9}$

62. $\dfrac{4}{9}$

63. $(a+4)(a-3)$ or a^2+a-12

64. $\dfrac{18}{(m-1)(m+2)}$ or $\dfrac{18}{m^2+m-2}$

65. $\dfrac{2(z+4)}{z-3}$

66. $-\dfrac{y+8}{y-8}$ or $\dfrac{y+8}{8-y}$

67. $\dfrac{x+2}{x+3}$

68. 1

69. $\dfrac{s+3}{s+4}$

70. $\dfrac{a-3}{2a-3}$

7.1   Mixed Exercises

71. $\dfrac{q}{5r^2}$

72. $-\dfrac{3}{2}$

73. $-z$

74. r

75. $-\dfrac{7}{8}$

76. -3

77. $(t+1)(t+2)$ or t^2+3t+2

78. $-x$

79. $\dfrac{p-6}{p+3}$

80. $\dfrac{(z+4)^3}{z(z-16)}$

7.2   Adding and Subtracting Rational Expressions

Objective 1

1. $\dfrac{11}{x}$

2. $-\dfrac{3}{y^2}$

3. $\dfrac{18}{5t}$

4. $\dfrac{c-4}{5a}$

5. $\dfrac{4n-7}{m+3}$

6. $z+y$

7. $\dfrac{1}{r-s}$

8. 1

9. $\dfrac{1}{x-5}$

10. $\dfrac{1}{k+5}$

11. $\dfrac{1}{q-7}$

12. $-\dfrac{1}{a+b}$

Objective 2

13. $30m$

14. $150z$

15. $75x^2 y$

16. $t(t-1)$

17. $24(s+3)$

18. $5a(a+2)$

19. $(q-6)(q+6)^2$

20. $r-p$ or $p-r$

21. $r(r+4)(r+1)$

22. $n(3+n)(3-n)$

23. $(p-4)(p+4)^2$

24. $(2z-1)(z+4)(z-3)$

Objective 3

25. $\dfrac{4y+35}{7y}$

26. $\dfrac{18+3x}{2x}$

27. $\dfrac{3z-5}{5z}$

28. $\dfrac{a}{3}$

29. $\dfrac{2+m}{2}$

30. $\dfrac{6-2s}{s^2}$

31. $\dfrac{18t+20}{15t^2}$

32. $2r+2$

33. $\dfrac{3}{(x+1)(x-1)(x+2)}$

34. $\dfrac{3y+1}{y^2-16}$

35. $\dfrac{3}{a-2}$ or $-\dfrac{3}{2-a}$

36. $-\dfrac{1}{m-3}$ or $\dfrac{1}{3-m}$

37. $\dfrac{8r}{r+2s}$

38. $\dfrac{2(a^2+3ab+4b^2)}{(b+a)^2(3b+a)}$

7.2   Mixed Exercises

39. $m-2$ or $2-m$

40. $3(x+2)(x+3)$

41. $\dfrac{1}{x+2}$

42. $\dfrac{3}{r-6}$ or $-\dfrac{3}{6-r}$

43. $\dfrac{8ab+21}{6a^2b}$

44. $\dfrac{2q^2+4q+4}{q(q+2)}$

45. $\dfrac{4k+24}{(k-2)(k+2)}$

46. $\dfrac{8m-4n}{m^2-n^2}$

47. $\dfrac{2z^2-z+1}{(z-1)(z+1)^2}$

48. $\dfrac{3y-5}{2(y+2)(y-2)}$

7.3   Complex Fractions

Objective 1

1. $\dfrac{y}{x}$

2. $\dfrac{3}{20}$

3. $\dfrac{2}{k}$

4. $\dfrac{(z+t)^2}{tz}$

5. $\dfrac{r}{s}$

6. $\dfrac{(m+1)^2}{m-1}$

7. $\dfrac{q(q^2+q+1)}{(q+1)^2(q-1)}$

8. $\dfrac{a-b}{4}$

9. $\dfrac{1}{rs}$

Objective 2

10. $\dfrac{k}{3}$

11. $\dfrac{m}{2}$

12. $\dfrac{3}{x}$

13. $\dfrac{b}{a}$

14. $\dfrac{2s}{5}$

15. $\dfrac{p^2+1}{3-p^2}$

16. $\dfrac{t^2-2}{t^2+4}$

17. $\dfrac{x+1}{1-x}$

18. $\dfrac{r-1}{r+1}$

Objective 3

19. n

20. $\dfrac{a^2+1}{1-a^2}$

21. $\dfrac{(z-y)^2}{z(z+y)}$

22. $\dfrac{-t+1}{t+1}$

23. $\dfrac{7(5k-m)}{4}$ or $\dfrac{35k-7m}{4}$

24. $\dfrac{1-r}{r(1+r)}$

Objective 4

25. $\dfrac{1}{5x+1}$

26. $\dfrac{x^2+xy+x+y}{x}$

27. $\dfrac{y^2+x}{xy^2}$

28. $\dfrac{2z^3+xy^2z^3}{x}$

29. $\dfrac{y^3}{4y^3z^2+z^2}$

30. $\dfrac{x^2y^2}{y^2-x^2}$

31. $\dfrac{2y^3}{x^2y^3+3x^2}$

32. $\dfrac{r}{s}$

33. $\dfrac{1}{xy-1}$

34. $\dfrac{m^2n^2}{(m+n)^2(n^2-m^2)}$

7.3   Mixed Exercises

35. $\dfrac{2x-1}{-x-1}$ or $-\dfrac{2x-1}{x+1}$

36. $\dfrac{1}{p}$

37. $\dfrac{n}{m(n+1)}$

38. $\dfrac{n(3n-2)}{3n-10}$

39. $\dfrac{m-2}{8}$

40. $\dfrac{s(s-1)}{2}$

41. $\dfrac{y-x}{x+y}$

42. $\dfrac{x^4}{4x^4+1}$

43. $1-x$

44. $\dfrac{y-x}{x^2y^2}$

7.4   Equations Involving Rational Expressions

Objective 1

1. (a) $0, -1$

    (b) $\{x \mid x \neq 0, -1\}$

2. (a) $\dfrac{5}{2}, -\dfrac{1}{3}$

    (b) $\left\{x \mid x \neq \dfrac{5}{2}, -\dfrac{1}{3}\right\}$

3. (a) $7, -8$

    (b) $\{x \mid x \neq 7, -8\}$

4. (a) -1

    (b) $\{x \mid x \neq -1\}$

5. (a) $0, 2$

    (b) $\{x \mid x \neq 0, 2\}$

6. (a) $-3, 0, 1$

    (b) $\{x \mid x \neq -3, 0, 1\}$

Objective 2

7. $\left\{\dfrac{11}{4}\right\}$

8. $\{-7\}$

9. $\{1\}$

10. $\{5\}$

11. $\{-6\}$

12. $\{-8\}$

13. $\{2\}$

14. $\{4, 6\}$

15. $\{3\}$

16. $\left\{-\dfrac{4}{3}, 1\right\}$

Objective 3

17.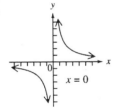

18.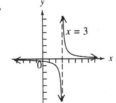

19.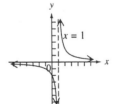

20.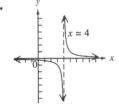

7.4   Mixed Exercises

21. $\left\{\dfrac{1}{3}\right\}$

22. $\left\{\dfrac{64}{3}\right\}$

23. $\{-3\}$

24. $\{-2\}$

25. $\varnothing$

26. $\{-6\}$

27. $\left\{\dfrac{4}{3}\right\}$

28. $\{2, 9\}$

29. $\left\{-6, \dfrac{1}{2}\right\}$

30. $\varnothing$

7.5   Applications of Rational Expressions

Objective 1

1. $b = \dfrac{5}{9}$

2. $R = \dfrac{1}{3}$

3. $m = \dfrac{75}{8}$

4. $L = 20$

5. $R_2 = 20$

6. $d_i = \dfrac{50}{3}$

7. $t = 2$

8. $r = \dfrac{1}{20}$ or .05 or 5%

9. $B = 7$

10. $h = 12$

Objective 2

11. $R_2 = \dfrac{RR_1}{R_1 - R}$

12. $d_0 = \dfrac{fd_i}{d_i - f}$

13. $T_2 = \dfrac{T_1 V_2 P_2}{V_1 P_1}$

14. $m_2 = \dfrac{Fd^2}{Gm_1}$

15. $a_n = \dfrac{2s_n}{n} - a_1$

16. $b_1 = \dfrac{2A}{h} - b_2$

17. $V_1 = \dfrac{T_1 V_2 P_2}{P_1 T_2}$

18. $R_1 = \dfrac{AR_2}{R_2 - A}$

19. $E = \dfrac{nE - Inr}{I}$

20. $r = \dfrac{eR}{E - e}$

Objective 3

21. 20 questions

22. 1750 crimes

23. $\dfrac{2}{9}$ job/hr

24. $38.40

25. $1125

26. 5 hr

27. 8 gal

28. $3.90

29. 50,000

30. $1000

Objective 4

31. 30 mi

32. 2 mph

33. 16 mi

34. 8 mph

35. 60 mph

36. Pauline: 60 mph; Pete: 40 mph

37. 60 mi

38. 12 km/hr

39. 120 mi

40. Ted: 4 mph; Olivia: 8 mph

Objective 5

41. $\dfrac{12}{7}$ or $1\dfrac{5}{7}$ hr

42. $\dfrac{9}{4}$ or $2\dfrac{1}{4}$ hr

43. $\dfrac{88}{19}$ or $4\dfrac{12}{19}$ hr

44. $\dfrac{143}{24}$ or $5\dfrac{23}{24}$ min

45. 60 hr

46. 60 hr

47. 24 hr

48. $\dfrac{15}{8}$ or $1\dfrac{7}{8}$ hr

49. $\dfrac{4}{5}$ hr or 48 min

50. $\dfrac{30}{11}$ or $2\dfrac{8}{11}$ hr

7.5   Mixed Exercises

51. $C = 37$

52. $b = 12$

53. $z = -\dfrac{15}{2}$

54. $p = .8$

55. $F = \dfrac{9C}{5} + 32$

56. $x = \dfrac{48yz}{35z + 120y}$

57. $r = \dfrac{56p}{8 + p}$

58. $B = \dfrac{2A - hb}{h}$

59. $v_s = \dfrac{Fv - fv - fv_0}{F}$

60. $v_0 = \dfrac{Fv - Fv_s - fv}{f}$

61. $\dfrac{20}{9}$ or $2\dfrac{2}{9}$ hr

62. 4 mph

63. 20 baskets

64. 36 hr

65. 14 mph

66. $3.84

67. 4 mph

68. $188.10

69. 275 mph

70. 2 hr

7.6 Variation

Objectives 1 and 2

1. $y = 3x$

2. $y = 5x$

3. $y = \dfrac{3x}{2}$

4. $y = \dfrac{x}{2}$

5. $y = \dfrac{23x}{12}$

6. $y = \dfrac{14x}{9}$

7. $y = 50x$

8. $y = 30x$

9. $y = \dfrac{4x}{3}$

10. $y = .25x$

11. 45

12. 144

13. $\dfrac{63}{2}$

14. $\dfrac{165}{4}$

15. 125

16. 147

17. 18

18. $\dfrac{49}{4}$

19. 69.08 cm

20. 100 psi

21. 100 newtons

22. 36π in.^{2}

Objective 3

23. $y = \dfrac{20}{x}$

24. $y = \dfrac{24}{x}$

25. $y = \dfrac{80}{x}$

26. $y = \dfrac{24}{x}$

27. $y = \dfrac{5}{6x}$

28. $y = \dfrac{27}{2x}$

29. $\dfrac{5}{2}$

30. 3

31. $\dfrac{4}{9}$

32. $\dfrac{64}{25}$

33. 100 amps

34. $\dfrac{400}{27}$ footcandles

35. 40 lb

36. 90 revolutions/min

Objective 4

37. $y = xz$

38. $y = \dfrac{25xz}{12}$

39. $y = \dfrac{xz}{2}$

40. $y = 2xz$

41. $y = \dfrac{xz}{2}$

42. $y = \dfrac{xz}{4}$

43. 24

44. 96

45. 128

46. 96

47. $800

48. 750°

Objective 5

49. $y = \dfrac{3x}{z}$

50. $y = \dfrac{12x}{z}$

51. $y = \dfrac{2x}{3z}$

52. $y = \dfrac{x}{3z}$

53. $y = \dfrac{.65x}{z}$

54. $y = \dfrac{3.5x}{z}$

55. 9 hr

56. 6,000,000 dynes

7.6 Mixed Exercises

57. $p = 7q$

58. $w = .02vs$

59. $\dfrac{14}{3}$

60. 256.25

61. $\dfrac{128}{81}$

62. 120

63. 144 ft

64. 850 ohms

65. 1.105 L

66. 20 lb

67. 12 km

68. $810

Chapter 8

LINEAR EQUATIONS AND APPLICATIONS

8.1   Review of Solving Linear Equations and Inequalities

Objective 1

1. Yes

2. No

3. Yes

4. Yes

5. Yes

6. No

7. Yes

8. No

9. Yes

10. Yes

11. $\{-9\}$

12. $\{-3\}$

13. $\{-1\}$

14. $\varnothing$

15. $\left\{-\dfrac{2}{7}\right\}$

16. $\{-5\}$

17. $\{1\}$

18. $\{8\}$

19. $\left\{-\dfrac{8}{3}\right\}$

20. $\left\{\dfrac{3}{2}\right\}$

21. $\{10\}$

22. $\{-2\}$

23. $\left\{\dfrac{19}{5}\right\}$

24. $\left\{\dfrac{5}{9}\right\}$

25. $\{0\}$

26. $\{0\}$

27. $\{4\}$

28. $\{-8\}$

29. $\{-4\}$

30. $\left\{\dfrac{2}{5}\right\}$

31. $\left\{-\dfrac{3}{4}\right\}$

32. $\left\{\dfrac{24}{19}\right\}$

33. $\{-1\}$

34. $\left\{-\dfrac{163}{25}\right\}$

35. $\{51\}$

36. $\left\{-\dfrac{98}{5}\right\}$

37. $\{-10\}$

38. $\{6\}$

39. $\left\{-\dfrac{3}{2}\right\}$

40. $\{-14\}$

41. $\{28\}$

42. $\left\{-\dfrac{108}{5}\right\}$

43. $\{2\}$

44. $\left\{-\dfrac{3}{2}\right\}$

45. $\{-5\}$

46. $\{3\}$

47. $\{300\}$

48. $\{60\}$

49. $\{210\}$

50. $\{21\}$

51. $\{47\}$

52. $\{146\}$

Objective 2

53. Conditional equation; {2}

54. Conditional equation; {6}

55. Contradiction; ∅

56. Identity; {All real numbers}

57. Contradiction; ∅

58. Identity; {All real numbers}

59. Conditional equation; {0}

60. Contradiction; ∅

61. Identity; {All real numbers}

62. Conditional equation; {1}

Objective 3

63. $(-\infty, -3]$

64. $(9, \infty)$

65. $(-6, \infty)$

66. $[9, \infty)$

67. $(-\infty, 0)$

68. $(3, \infty)$

69. $(-\infty, -3)$

70. $[-2, \infty)$

71. $(1, \infty)$

72. $(-\infty, 15]$

73. $(4, \infty)$

74. $(6, \infty)$

75. $(-\infty, 3]$

76. $(-3, \infty)$

77. $(-\infty, -4)$

78. $\left(-\infty, -\dfrac{5}{2}\right]$

79. $[4, \infty)$

80. $(-\infty, -10]$

81. $(-\infty, -36]$

82. $[10, \infty)$

83. $(-\infty, 14)$

84. $(-\infty, -2]$

85. $(-\infty, -4)$

86. $(-8, \infty)$

87. $[-3, \infty)$

Objective 4

88. $(1, 9)$

89. $(-8, 6)$

90. $(5, 9)$

91. $[-9, -7)$

92. $(4, 5)$

93. $\left(-\dfrac{8}{3}, 0\right)$

94. $\left[-\dfrac{19}{2}, \dfrac{29}{2}\right]$

95. $\left[-\dfrac{11}{6}, \dfrac{1}{6}\right]$

96. $[-8, 12]$

97. $[-15, 10]$

8.1   Mixed Exercises

98. No

99. Yes

100. $\{-3\}$

101. $\{11\}$

102. $\{-7\}$

103. $\left\{-\dfrac{5}{2}\right\}$

104. $\left\{\dfrac{14}{5}\right\}$

105. {All real numbers}

106. $\{12\}$

107. $\{-21\}$

108. $\{0\}$

109. $\left\{-\dfrac{24}{31}\right\}$

110. $\varnothing$

111. {All real numbers}

112. $\{8\}$

113. $\{6\}$

114. $\left(-\dfrac{3}{2}, \infty\right)$

115. $\left[\dfrac{1}{2}, \dfrac{25}{2}\right]$

116. $\left(-\dfrac{1}{3}, \dfrac{2}{3}\right)$

117. $\left(-\dfrac{7}{2}, 1\right)$

118. $\left[-1, \dfrac{13}{3}\right]$

119. $\left(-\infty, -\dfrac{3}{8}\right]$

8.2 Set Operations and Compound Inequalities

Objective 1

1. $\{2, 3\}$ **3.** $\varnothing$ **5.** $\{2, 4\}$ **7.** $\{0, 2, 4\}$ or D **9.** $\varnothing$

2. $\varnothing$ **4.** $\{2, 6, 8\}$ **6.** $\{1, 3, 5\}$ **8.** $\{0\}$ or E **10.** $\{2, 4\}$

Objective 2

11. $(0, 3)$

12. $(-\infty, 4]$

13. $[1, \infty)$

14. $[5, 9]$

15. $(-2, -1)$

16. $\varnothing$

17. $[-4, 3)$

18. $(0, 1]$

19. $\varnothing$

20. $\varnothing$

Objective 3

21. $\{0, 1, 2, 3, 4, 5\}$

22. $\{-6, -5, -4, -3, -2, -1\}$

23. $\{7, 8, 9, 10\}$

24. $\{2, 6, 8\}$

25. $\{0, 1, 2, 3, 4, 5, 6, 8, 10\}$

26. $\{1, 2, 3, 4, 5, 6, 7, 9\}$

27. $\{1, 2, 3, 4, 5, 6\}$ or A

28. $\{0, 1, 2, 3, 4, 5, 6\}$

29. $\{0, 1, 2, 3, 4, 5, 6, 7, 8, 9, 10\}$

30. $\{0, 1, 2, 3, 4, 6, 8, 10\}$

Objective 4

31. $(-\infty, -1) \cup (4, \infty)$

32. $(-\infty, 1] \cup [6, \infty)$

33. $(-\infty, 2] \cup [6, \infty)$

34. $(-\infty, 2] \cup [6, \infty)$

35. $[-1, \infty)$

36. $(-\infty, \infty)$

37. $(-\infty, -4] \cup (4, \infty)$

38. $(-\infty, -1) \cup (5, \infty)$

39. $\left(-\infty, -\dfrac{5}{3}\right) \cup \left(\dfrac{3}{4}, \infty\right)$

40. $(-\infty, \infty)$

8.2 Mixed Exercises

41. $\{0, 2, 4\}$

42. $\{1, 3, 5\}$

43. $\{0, 1, 2, 3, 4, 5\}$ or D

44. $\varnothing$

45. $(-\infty, -2] \cup [0, \infty)$

46. $[-2, 3)$

47. $(-\infty, 1) \cup [2, \infty)$

48. $(-\infty, 5) \cup (6, \infty)$

49. $\varnothing$

50. $(-\infty, \infty)$

51. $(-1, 4]$

52. $\left(-\infty, -\dfrac{2}{7}\right]$

8.3   Absolute Value Equations and Inequalities

Objective 1

1. -7 7

2. 0

3. -8 8

4. -2 2

5. -6 6

6.

7. -2 0

8. -3 3

9. -10 10

10. 0

Objective 2

11. $\{-3, 11\}$

12. $\{-12, 4\}$

13. $\{-4, 10\}$

14. $\left\{-\dfrac{5}{3}, \dfrac{7}{3}\right\}$

15. $\{2, 8\}$

16. $\{-8, -4\}$

17. $\left\{-\dfrac{13}{2}, \dfrac{7}{2}\right\}$

18. $\left\{-\dfrac{3}{2}\right\}$

19. $\{3\}$

20. $\left\{-3, \dfrac{21}{2}\right\}$

Objective 3

21. $(-\infty, -6) \cup (10, \infty)$
-6 10

22. $(-\infty, -20) \cup (10, \infty)$
-20 10

23. $(-13, 3)$
-13 3

24. $(-3, 13)$
-3 13

25. $\left(-4, \dfrac{16}{5}\right)$
-4 $\dfrac{16}{5}$

26. $(-\infty, -7] \cup [16, \infty)$
-7 16

27. $\left(-\infty, \dfrac{1}{3}\right] \cup [3, \infty)$
$\dfrac{1}{3}$ 3

28. $(-\infty, -4] \cup [-2, \infty)$
-4 -2

29. $(-\infty, -3) \cup (-1, \infty)$
-3 -1

30. $(-\infty, 0] \cup [10, \infty)$
0 10

31. $[-1, 5]$
-1 5

32. $\left[-\dfrac{1}{4}, \dfrac{3}{4}\right]$
$-\dfrac{1}{4}$ $\dfrac{3}{4}$

Objective 4

33. $\{-2, 2\}$

34. $\{-9, 9\}$

35. $\varnothing$

36. $\varnothing$

37. $\{-9, -1\}$

38. $\left\{-\dfrac{13}{7}, \dfrac{3}{7}\right\}$

39. $\varnothing$

40. $\{-2, 3\}$

41. $\{-42, 50\}$

42. $\left\{-\dfrac{5}{4}, -\dfrac{1}{4}\right\}$

Objective 5

43. $\left\{\dfrac{3}{5}, 15\right\}$

44. $\left\{\dfrac{7}{2}\right\}$

45. $\left\{\dfrac{2}{3}, 8\right\}$

46. $\left\{-8, \dfrac{2}{3}\right\}$

47. $\{-1\}$

48. $\left\{-\dfrac{3}{2}, 2\right\}$

49. $\left\{-\dfrac{15}{4}, \dfrac{1}{8}\right\}$

50. $\left\{-3, \dfrac{11}{3}\right\}$

51. $\{-1, 1\}$

52. $\left\{-\dfrac{1}{2}, \dfrac{5}{2}\right\}$

Objective 6

53. $\varnothing$

54. $\{-14\}$

55. $\{0\}$

56. $\varnothing$

57. $\varnothing$

58. $\varnothing$

59. $\varnothing$

60. $\varnothing$

61. $(-\infty, \infty)$

62. $(-\infty, \infty)$

8.3 Mixed Exercises

63. $\{-4, -1\}$

64. $\{-2, 14\}$

65. $\varnothing$

66. $\{-11, 1\}$

67. $\varnothing$

68. $\left\{-\dfrac{29}{3}, \dfrac{25}{3}\right\}$

69. $\left\{-5, -\dfrac{3}{4}\right\}$

70. $\{-6, 0\}$

71. $(-2, 5)$

72. $\left[-\dfrac{1}{3}, \dfrac{19}{3}\right]$

73. $(-\infty, -8) \cup (1, \infty)$

74. $(-\infty, -4] \cup [10, \infty)$

75. $(-\infty, -6] \cup [5, \infty)$

76. $\left(-\infty, -\dfrac{1}{2}\right) \cup \left(\dfrac{1}{2}, \infty\right)$

77. $\left[-\dfrac{1}{3}, 1\right]$

78. $\left(-\dfrac{1}{2}, 2\right)$

8.4   Systems of Linear Equations in Two Variables

Objective 1

1. $\{(1, 2)\}$

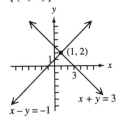

2. $\{(4, 1)\}$

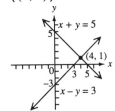

3. $\{(-2, 0)\}$

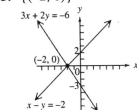

4. $\{(1, 0)\}$

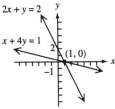

5. $\{(-2, 3)\}$

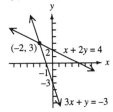

6. $\{(3, 1)\}$

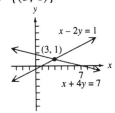

7. $\{(-4, 2)\}$

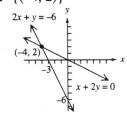

8. $\{(0, 2)\}$

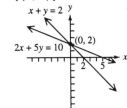

9. solution

10. solution

11. not a solution

12. not a solution

13. not a solution

14. solution

15. solution

16. solution

17. not a solution

18. not a solution

19. $\{(-4, -8)\}$

20. $\{(-1, 3)\}$

21. $\{(3, 1)\}$

22. $\{(3, -1)\}$

23. $\{(3, 1)\}$

24. $\{(-1, 8)\}$

25. $\{(3, 5)\}$

26. $\left\{\left(-\dfrac{1}{2}, 4\right)\right\}$

27. $\{(3, 2)\}$

28. $\{(-3, -4)\}$

29. $\{(1, 3)\}$

30. $\{(-1, -5)\}$

31. $\{(-1, 5)\}$

32. $\{(2, 1)\}$

33. $\{(3, 2)\}$

34. $\{(1, 3)\}$

35. $\{(4, 1)\}$

36. $\{(4, -2)\}$

37. $\{(-3, -1)\}$

38. $\{(-1, 7)\}$

39. $\left\{\left(\dfrac{1}{2}, \dfrac{1}{4}\right)\right\}$

40. $\left\{\left(\dfrac{2}{3}, \dfrac{1}{4}\right)\right\}$

41. $\{(6, 8)\}$

42. $\{(-5, 2)\}$

43. $\{(-2, -3)\}$

Objective 2

44. $\varnothing$

45. $\varnothing$

46. $\{(x, y) \mid 5x - 2y = 4\}$

47. $\{(x, y) \mid 9x - y = 6\}$

48. $\varnothing$

8.4 Mixed Exercises

49. $\{(3, 4)\}$

50. $\{(3, -1)\}$

51. $\{(0, -4)\}$

52. $\{(3, -1)\}$

53. $\{(x, y) \mid 4x + 6y = 24\}$

54. $\{(6, 2)\}$

55. $\{(2, -1)\}$

56. $\{(-3, 0)\}$

57. $\{(2, 4)\}$

58. $\varnothing$

59. $\varnothing$

60. $\{(x, y) \mid -x + 2y = -8\}$

8.5   Systems of Linear Equations in Three Variables

Objective 2

1. $\{(-2, 1, 1)\}$

2. $\{(1, 0, 3)\}$

3. $\{(3, -1, 2)\}$

4. $\{(-3, 1, 2)\}$

5. $\{(0, -2, 5)\}$

6. $\{(3, 0, -2)\}$

7. $\{(4, -4, 1)\}$

8. $\{(3, -6, 1)\}$

9. $\{(-12, 18, 0)\}$

10. $\{(-15, 0, 16)\}$

Objective 3

11. $\{(2, -1, 5)\}$

12. $\left\{ \left(\dfrac{1}{2}, \dfrac{2}{3}, \dfrac{1}{5} \right) \right\}$

13. $\{(2, -5, 3)\}$

14. $\{(-1, -1, 5)\}$

15. $\{(3, 5, 0)\}$

16. $\{(0, -2, 5)\}$

17. $\{(4, 2, -1)\}$

18. $\{(2, -3, 1)\}$

19. $\{(6, -4, 1)\}$

20. $\{(-3, 5, -6)\}$

Objective 4

21. $\varnothing$

22. $\varnothing$

23. $\{(x, y, z) \mid 3x - 2y + 4z = 5\}$

24. $\{(x, y, z) \mid -x + 5y - 2x = 3\}$

25. $\{(0, 0, 0)\}$

26. $\{(0, 0, 0)\}$

27. $\varnothing$

28. $\varnothing$

29. $\{(x, y, z) \mid x - 5y + 2z = 0\}$

30. $\{(x, y, z) \mid 3x - 2y + 5z = 0\}$

31. 9, 10, 12

32. 28, 31, 40

33. 50°, 60°, 70°

34. 40°, 65°, 75°

35. 11 in., 21 in., 28 in.

36. 7 fives, 12 tens,
    32 twenties

37. 12 tens, 25 twenties, 13 fifties

38. $40,000 at 5%; $10,000 at 6%; $30,000 at 7%

39. $20,000 at 8%; $50,000 at 10%; $5000 at 11%

40. 15 lb of $8; 12 lb of $10; 23 lb of $15

8.5   Mixed Exercises

41. $\{(1, 2, 3)\}$

42. $\{(3, 2, 1)\}$

43. $\{(0, 5, -3)\}$

44. $\{(x, y, z) \mid -3x - y + 2z = -3\}$

45. $\{(2, 2, 5)\}$

46. $\varnothing$

47. $\{(2, -1, 0)\}$

48. $\{(2, -1, 3)\}$

49. $\{(-2, -4, 0)\}$

50. $\{(1, 1, 2)\}$

51. side of square: 8 cm;
side of triangle: 13 cm

8.6   Solving Systems of Linear Equations by Matrix Methods

Objective 1

1. 2×2

2. 3×2

3. 2×3

4. 4×2

5. 4×3

6. 3×4

Objective 2

7. $\begin{bmatrix} 3 & -4 & | & 7 \\ 2 & 1 & | & 12 \end{bmatrix}$

8. $\begin{bmatrix} 2 & -3 & | & 12 \\ 7 & 3 & | & 15 \end{bmatrix}$

9. $\begin{bmatrix} \frac{1}{3} & -\frac{1}{2} & | & 7 \\ \frac{5}{3} & \frac{1}{2} & | & 8 \end{bmatrix}$

10. $\begin{bmatrix} \frac{1}{2} & \frac{1}{2} & | & -16 \\ -3 & 1 & | & 2 \end{bmatrix}$

11. $\begin{bmatrix} 3 & -1 & | & 4 \\ -5 & 1 & | & -2 \end{bmatrix}$

12. $\begin{bmatrix} 2 & 1 & | & -3 \\ 1 & -4 & | & -5 \end{bmatrix}$

13. $\begin{bmatrix} -2 & 3 & -5 & | & 7 \\ 6 & 2 & -4 & | & 12 \\ 5 & -2 & 1 & | & -1 \end{bmatrix}$

14. $\begin{bmatrix} 1 & 1 & 1 & | & 10 \\ 2 & 1 & -3 & | & 11 \\ 1 & 0 & 2 & | & -2 \end{bmatrix}$

Objective 3

15. $\{(2, -3)\}$

16. $\{(3, 2)\}$

17. $\{(3, -2)\}$

18. $\{(0, -2)\}$

19. $\{(0, 3)\}$

20. $\{(1, -1)\}$

21. $\{(3, 1)\}$

22. $\{(-5, -3)\}$

23. $\left\{ \left(3, \frac{3}{2} \right) \right\}$

Objective 4

24. $\{(2, -1, -3)\}$

25. $\{(2, -1, 4)\}$

26. $\{(1, 3, -2)\}$

27. $\{(3, -1, 4)\}$

28. $\{(1, 2, 4)\}$

29. $\{(-2, 1, 2)\}$

Objective 5

30. $\varnothing$

31. $\varnothing$

32. $\{(x, y) \mid x - 2y = 3\}$

33. $\{(x, y) \mid 2x + y = 10\}$

34. $\varnothing$

35. $\varnothing$

36. $\{(1, 0, 0)\}$

37. $\{(x, y, z) \mid x + 2y - z = 6\}$

8.6   Mixed Exercises

38. 2×3

39. 4×3

40. $\begin{bmatrix} 2 & -5 & | & 2 \\ 3 & 4 & | & -1 \end{bmatrix}$

41. $\begin{bmatrix} 2 & 1 & | & 7 \\ 1 & -4 & | & 0 \end{bmatrix}$

42. $\varnothing$

43. $\{(-2, -3)\}$

44. $\{(4, -2)\}$

45. $\left\{ (x, y) \mid 2x - y = 2 \right\}$

46. $\varnothing$

47. $\{(1, -1, -1)\}$

48. $\{(-4, 3, 2)\}$

Chapter 9

ROOTS AND RADICALS

9.1   Radical Expressions and Graphs

Objective 1

1. 3	**6.** 3	**11.** 15	**16.** −7
2. 6	**7.** −5	**12.** −4	**17.** 9
3. −2	**8.** 2	**13.** 2	**18.** 8
4. 2	**9.** 5	**14.** 3	**19.** −11
5. not a real number	**10.** −1	**15.** 4	**20.** 11

Objective 3

21. 3

22. 4

23. −3

24. 1

25. −3

26. $\left| x^3 \right|$

27. $\left| x^3 \right|$

28. y^2

29. $-a^3$

30. −2

31. r^2

32. $-c^2$

33. 5

34. c^9

35. −2

36. a^2

37. 2

38. −2

39. −1

40. 7

41. −12

42. $-\dfrac{7}{4}$

43. $-\dfrac{11}{5}$

44. 5

45. $\left| t \right|$

46. s^8

47. p^{10}

48. a^{18}

Objective 4

49.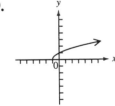

domain: $[-1, \infty)$
range: $[0, \infty)$

50.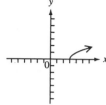

domain: $[3, \infty)$
range: $[0, \infty)$

51.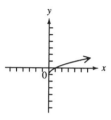

domain: $[0, \infty)$
range: $[-1, \infty)$

52.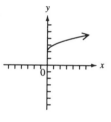

domain: $[0, \infty)$
range: $[2, \infty)$

53.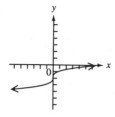

domain: $(-\infty, \infty)$
range: $(-\infty, \infty)$

54.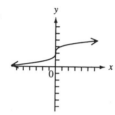

domain: $(-\infty, \infty)$
range: $(-\infty, \infty)$

Objective 5

55. $|x|$

56. $-|x|$

57. x

58. $-x$

59. x^2

60. $-x^2$

61. $-x^4$

62. x^3

Objective 6

63. 9.327

64. 8.718

65. −3.936

66. −3.271

67. 2.546

68. 3.027

69. 10.050

70. −12.610

71. 2.897

72. 6.407

73. −17.607

74. 2.418

75. 8.707

76. 29.496

77. 30.496

78. −3.377

79. −19.235

80. −3.973

81. 11.091

82. 3.037

9.1   Mixed Exercises

83. 14

84. 8

85. −15

86. 19.596

87. −18.248

88. 5.180

89. k^2

90. $|d|$

91. $-p^2$

92. 3.503

93. not a real number

94. 2.590

9.2 Rational Exponents

Objective 1

1. 3

2. 5

3. 2

4. 7

5. −3

6. not a real number

7. 10

8. −2

9. 5

10. 2

11. −4

12. 2

13. 6

14. 3

15. 15

16. 8

17. 9

18. 5

19. 6

20. −5

Objective 2

21. 9

22. 125

23. 243

24. $\dfrac{1}{9}$

25. $\dfrac{1}{125}$

26. $\dfrac{1}{9}$

27. 100

28. $-\dfrac{1}{25}$

29. 8

30. $\dfrac{1}{36}$

31. $\dfrac{1}{27}$

32. $\dfrac{1}{3}$

33. 8

34. $-\dfrac{1}{4}$

35. $\dfrac{1}{4}$

36. −125

37. $\dfrac{1}{64}$

38. 7776

39. 243

40. $\dfrac{1}{2}$

Objective 3

41. x^6

42. $\sqrt[3]{2t^2}$

43. $\sqrt[6]{x^5}$

44. 7^2 or 49

45. $x\sqrt{2x}$

46. $\sqrt[8]{y^7}$

47. $\sqrt[4]{5^3}$

48. $\sqrt[12]{2^7}$

49. $\sqrt[6]{x^5}$

50. $p\sqrt[5]{p^2}$

51. $\sqrt[6]{r}$

52. $\sqrt[5]{3t^2}$

53. $\sqrt[6]{k^5}$

54. $\sqrt[15]{p^{11}}$

55. $r^2\sqrt[3]{r}$

56. $x\sqrt[12]{x}$

57. $\sqrt[12]{x^4 y^3 z^6}$

58. $\sqrt[4]{20}$

Objective 4

59. 13^2 or 169

60. $\dfrac{1}{5^{1/7}}$

61. 8

62. 7^4 or 2401

63. 3^2 or 9

64. 5^3 or 125

65. $r^{11/12}$

66. y

67. $a^{5/6}$

68. a^4

69. 5^8 or 390,625

70. a

71. $x^{19/12}$

72. 8

73. $a^{2/15}$

74. $\dfrac{c^4}{x^2}$

75. $\dfrac{y^{10/3}}{x^{14/5}}$

76. $\dfrac{y}{x^{11/3}}$

9.2   Mixed Exercises

77. -3

78. 2

79. $-\dfrac{8}{125}$

80. $\dfrac{64}{27}$

81. -100

82. $-\dfrac{1}{8}$

83. $\dfrac{1}{125}$

84. 100

85. 1

86. x^2

87. $x\sqrt[9]{x}$

88. $\sqrt[15]{x}$

9.3   Simplifying Radical Expressions

Objective 1

1. $\sqrt{33}$

2. $\sqrt{70}$

3. $\sqrt{2t}$

4. $\sqrt{42xt}$

5. $\sqrt{\dfrac{91}{cw}}$

6. $\sqrt{\dfrac{33}{rp}}$

7. $\sqrt[4]{42}$

8. $\sqrt[6]{20t^5}$

9. $\sqrt[5]{24r^4t^4}$

10. $\sqrt[3]{21}$

11. $\sqrt{35}$

12. $\sqrt{165}$

13. $\sqrt{6r}$

14. $\sqrt[3]{5x}$

15. $\sqrt[3]{35xy}$

16. $\sqrt[4]{120}$

17. $\sqrt[7]{48a^3t^6}$

18. $\sqrt[4]{4x}$

Objective 2

19. $\dfrac{5}{4}$

20. $\dfrac{\sqrt{5}}{6}$

21. $\dfrac{\sqrt{x}}{9}$

22. $\dfrac{t^4\sqrt{t}}{5}$

23. $-\dfrac{3}{2}$

24. $\dfrac{\sqrt[3]{45}}{3}$

25. $\dfrac{7}{10}$

26. $\dfrac{\sqrt{6}}{7}$

27. $\dfrac{5}{r^5}$

28. $-\dfrac{a^2}{5}$

29. $\dfrac{w}{6}$

30. $\dfrac{\sqrt[3]{\ell^2}}{3}$

31. $\dfrac{\sqrt{r}}{11}$

32. $\dfrac{\sqrt[4]{p}}{2}$

33. $-\dfrac{7}{5}$

34. $\dfrac{z^2}{6}$

35. $\dfrac{\sqrt[5]{7x}}{2}$

36. $\dfrac{\sqrt{15}}{13}$

Objective 3

37. $3\sqrt{3}$

38. $3\sqrt{7}$

39. $5\sqrt{3}$

40. $10\sqrt{2}$

41. $3\sqrt{15}$

42. $2\sqrt[3]{3}$

43. $2\sqrt[4]{2}$

44. $6t^2\sqrt{t}$

45. $5r^2\sqrt{2rx}$

46. $2y^2\sqrt[3]{3}$

47. $3x^3\sqrt[3]{2x^2}$

48. $2x^3c^2\sqrt[3]{10c}$

49. $3a\sqrt[3]{4a^2}$

50. $7r^4\sqrt{r}$

51. $3bc^2\sqrt[3]{10bc^2}$

Objective 4

52. $x\sqrt[3]{x}$

53. $x\sqrt{x}$

54. $\sqrt[6]{5}$

55. $x^2\sqrt{x}$

56. $\sqrt[3]{9}$

57. $\sqrt[3]{x^2}$

58. $z\sqrt[4]{z}$

59. $\sqrt[4]{x^3y^2}$

60. $\sqrt[3]{z^2 y}$ **61.** $\sqrt[4]{9x^2 y^3}$ **62.** $\sqrt[6]{5400}$ **63.** $\sqrt[6]{63}$

Objective 5

64. 10

65. 26

66. $2\sqrt{2}$

67. $2\sqrt{6}$

68. $\sqrt{13}$

69. $6\sqrt{2}$

70. $4\sqrt{5}$

71. $4\sqrt{6}$

72. $\sqrt{39}$

73. $3\sqrt{5}$

Objective 6

74. $2\sqrt{5}$

75. $2\sqrt{13}$

76. $\sqrt{10}$

77. $\sqrt{61}$

78. $\sqrt{34}$

79. $5\sqrt{2}$

80. $\sqrt{58}$

81. $\sqrt{181}$

82. $\sqrt{37}$

83. $4\sqrt{3}$

84. $\sqrt{x^2 + 4y^2}$

85. $\sqrt{a^2 - 2ab + 5b^2}$

9.3   Mixed Exercises

86. $\sqrt{30ab}$

87. $2x^3 y^2 \sqrt{y}$

88. $\sqrt[5]{8w^4}$

89. $3t^2 \sqrt[3]{2t}$

90. 5

91. $x^3 \sqrt{3xy}$

92. $\dfrac{3}{2}$

93. $\dfrac{a^2}{25}$

94. $\dfrac{b^4 q^2 \sqrt{b}}{2}$

95. $2xy^3 z^5 \sqrt{2xz}$

96. $v^3 \sqrt[4]{t^3}$

97. $2\sqrt[10]{972}$

98. 13

99. $\sqrt{105}$

100. $2\sqrt{17}$

101. 5

9.4   Adding and Subtracting Radical Expressions

Objective 1

1. $5\sqrt{7}$

2. $13\sqrt{13}$

3. $27\sqrt{3}$

4. $49\sqrt{2}$

5. $22\sqrt[3]{3}$

6. $5\sqrt[4]{2}$

7. $3\sqrt{13}$

8. $-15\sqrt{6}$

9. $9\sqrt{3y}$

10. $7\sqrt{2}$

11. $9\sqrt[3]{3}$

12. $6\sqrt[3]{5}$

13. $12\sqrt{x}$

14. $-4\sqrt[3]{3}$

15. $12\sqrt{2}$ in., 18 in.^{2}

16. $11\sqrt{5}$ cm

17. $10\sqrt{3}$ ft, 18 ft^2

18. $35\sqrt{15}$ cm^2

19. $9\sqrt{10}$ m^2

9.4   Mixed Exercises

20. $\sqrt{10}$

21. $-\sqrt{6}$

22. $-3\sqrt[3]{2}$

23. $24\sqrt[3]{5}$

24. $5\sqrt[3]{2r}$

25. $13\sqrt{2z}$

26. $13\sqrt{2z}$

27. $28z\sqrt[3]{2}$

28. $3\sqrt{3}$ cm, 135 cm^2

29. $2\sqrt{35}$ in.

9.5 Multiplying and Dividing Radical Expressions

Objective 1

1. $6 + 3\sqrt{7} + 2\sqrt{2} + \sqrt{14}$

2. $2\sqrt{15} - \sqrt{110} + 3\sqrt{2} - \sqrt{33}$

3. 23

4. $\sqrt{10} - 4\sqrt{5} + 2\sqrt{3} - 4\sqrt{6}$

5. -14

6. $52 + 30\sqrt{3}$

7. $-16 + 13\sqrt{2}$

8. $14 - 4\sqrt{6}$

9. 4

10. $x - y^2$

11. $6x - 13\sqrt{x} + 6$

12. $4 - \sqrt[3]{25}$

Objective 2

13. $\dfrac{6\sqrt{5}}{5}$

14. $3\sqrt{5}$

15. $\dfrac{\sqrt{22}}{11}$

16. $\dfrac{\sqrt{30}}{2}$

17. $\dfrac{\sqrt{2}}{2}$

18. $\dfrac{3\sqrt{2}}{4}$

19. $\dfrac{7\sqrt{3}}{15}$

20. $\dfrac{-14\sqrt{3}}{9}$

21. $\dfrac{6\sqrt{2}}{5}$

22. $\dfrac{2\sqrt{6}}{3}$

23. $\dfrac{7\sqrt{3}}{12}$

24. $\dfrac{\sqrt{15}}{40}$

25. $\dfrac{\sqrt{5}}{2}$

26. $\dfrac{\sqrt{10}}{2}$

27. $\dfrac{3\sqrt{2}}{14}$

28. $\dfrac{3}{4}$

29. $\dfrac{6\sqrt{t}}{t}$

30. $\dfrac{5\sqrt{2r}}{r}$

31. $\dfrac{9x^2\sqrt{2t}}{t^3}$

32. $\dfrac{2\sqrt{2m}}{m}$

33. $\dfrac{5\sqrt{y}}{y}$

34. $\dfrac{\sqrt{10x}}{4}$

35. $\dfrac{2x\sqrt{3}}{3}$

36. $\dfrac{3\sqrt{14z}}{7}$

37. $\dfrac{2\sqrt{7y}}{7y}$

38. $\dfrac{\sqrt{10}}{4x}$

39. $\dfrac{2a\sqrt{15t}}{5t^2}$

40. $\dfrac{y\sqrt{21b}}{6b}$

41. $\dfrac{\sqrt{38}}{8}$

42. $\dfrac{\sqrt{5}}{5}$

43. $\dfrac{\sqrt[3]{3}}{3}$

44. $\dfrac{\sqrt[3]{42}}{9}$

45. $\dfrac{\sqrt[3]{10}}{5}$

46. $\dfrac{\sqrt[3]{xy^2}}{y}$

47. $\dfrac{\sqrt[3]{3}}{2}$

48. $\dfrac{\sqrt[3]{20x}}{2}$

49. $\dfrac{\sqrt[3]{42}}{6}$

50. $\dfrac{\sqrt[3]{100}}{5}$

51. $\dfrac{c\sqrt[3]{d}}{d}$

52. $\dfrac{t^2\sqrt[3]{x^2}}{x^3}$

53. $\dfrac{\sqrt[3]{28r}}{14}$

54. $\dfrac{\sqrt[3]{4mx^2}}{2x}$

Objective 3

55. $\dfrac{5\left(7+\sqrt{3}\right)}{46}$

56. $3\left(\sqrt{7}-3\right)$

57. $\dfrac{5\left(3+\sqrt{5}\right)}{4}$

58. $\dfrac{3\left(6-2\sqrt{2}\right)}{7}$ or $\dfrac{6\left(3-\sqrt{2}\right)}{7}$

59. $2\left(\sqrt{11}-\sqrt{2}\right)$

60. $-2\left(\sqrt{7}+\sqrt{5}\right)$

61. $\dfrac{-5\left(\sqrt{3}-\sqrt{11}\right)}{8}$

62. $\dfrac{4-\sqrt{5}}{11}$

63. $\dfrac{4+\sqrt{7}}{9}$

64. $-2\left(\sqrt{2}+\sqrt{3}\right)$

65. $\dfrac{7\left(\sqrt{3}+1\right)}{2}$

66. $-2\left(\sqrt{6}+\sqrt{3}\right)$

67. $\dfrac{-5\left(\sqrt{5}+\sqrt{3}\right)}{2}$

68. $\dfrac{\sqrt{3}\left(\sqrt{5}+\sqrt{2}\right)}{3}$ or $\dfrac{\sqrt{15}+\sqrt{6}}{3}$

69. $\dfrac{\sqrt{6}\left(\sqrt{13}-\sqrt{5}\right)}{8}$ or $\dfrac{\sqrt{78}-\sqrt{30}}{8}$

Objective 4

70. $4-\sqrt{2}$

71. $\dfrac{6-9\sqrt{3}}{4}$

72. $\dfrac{3+2\sqrt{15}}{4}$

73. $5-\sqrt{6}$

74. $\dfrac{1-\sqrt{2}}{2}$

75. $\dfrac{1+2\sqrt{3}}{5}$

76. $2+3\sqrt{3}$

77. $\dfrac{1-\sqrt{2}}{2}$

78. $2-\sqrt{3}$

79. $\dfrac{25+2\sqrt{5x}}{5}$

80. $3+\sqrt{5}$

81. $\dfrac{2-9\sqrt{2}}{3}$

9.5   Mixed Exercises

82. $3\sqrt{6}-12$

83. $\sqrt[3]{4}-1$

84. $\dfrac{-15\sqrt{2}}{4}$

85. $\dfrac{4\sqrt{3}}{3}$

86. $\dfrac{3\sqrt{2t}}{t}$

87. $\dfrac{-3\sqrt{3}-21+\sqrt{21}+7\sqrt{7}}{46}$

88. $\dfrac{6\sqrt{tx}-3x}{4t-x}$

89. $\dfrac{t-3\sqrt{t}}{t-9}$

90. $\dfrac{3\sqrt{3t}}{t^2}$

91. $\dfrac{\sqrt{30}}{10}$

92. $\dfrac{\sqrt{c}}{x}$

93. $\dfrac{\sqrt{15ab}}{3b}$

94. $\dfrac{x\sqrt{y}}{y^2}$

95. $\dfrac{a\sqrt[3]{ab^2}}{b}$

96. $\dfrac{\sqrt[3]{36s^2}}{6}$

97. $\dfrac{t^5\sqrt[3]{x^2}}{x}$

98. $\dfrac{\sqrt[3]{18xz^2}}{3z}$

99. $\dfrac{4-2\sqrt{2}}{3}$

100. $\dfrac{1-y\sqrt{2y}}{2}$

101. $\dfrac{6+\sqrt{5}}{5}$

102. $1+2\sqrt{2}$

103. $1+\sqrt{2x}$

104. $\dfrac{\sqrt{5}-1}{7}$

9.6   Solving Equations with Radicals

Objective 1

1. $\{25\}$
2. $\{2\}$
3. $\{6\}$
4. $\{8\}$
5. $\{10\}$
6. $\{0\}$
7. $\{15\}$
8. $\{1\}$
9. $\{10\}$

10. $\{11\}$
11. $\{6\}$
12. $\{16\}$
13. $\{41\}$
14. $\{4\}$
15. $\{100\}$
16. $\{15\}$
17. $\{7\}$
18. $\{21\}$

19. $\{7\}$
20. $\{20\}$
21. $\varnothing$
22. $\{16\}$
23. $\{23\}$
24. $\varnothing$
25. $\{15\}$
26. $\{5\}$
27. $\varnothing$

28. $\{13\}$
29. $\{2\}$
30. $\{1\}$
31. $\{22\}$
32. $\{11\}$
33. $\{2\}$
34. $\{17\}$
35. $\varnothing$
36. $\{36\}$

Objective 2

37. $\{4\}$
38. $\{1\}$
39. $\{2\}$
40. $\{3, 4\}$
41. $\{10\}$
42. $\varnothing$
43. $\{6\}$
44. $\varnothing$
45. $\{6\}$
46. $\{6\}$

47. $\left\{\dfrac{3}{4}\right\}$
48. $\{3, 4\}$
49. $\{3\}$
50. $\{-1\}$
51. $\{3\}$
52. $\{0, 8\}$
53. $\{3\}$
54. $\left\{\dfrac{3}{2}\right\}$
55. $\{7\}$

56. $\{3\}$
57. $\{-1\}$
58. $\{3\}$
59. $\{-2\}$
60. $\{7\}$
61. $\{1\}$
62. $\{-4, 0\}$
63. $\{5\}$
64. $\{11\}$

65. $\{0, -1\}$
66. $\{-7, -2\}$
67. $\{6\}$
68. $\{-11, 13\}$
69. $\{3\}$
70. $\{9\}$
71. $\{2\}$
72. $\{-9\}$

Objective 3

73. $\{3\}$
74. $\{28\}$
75. $\{-3\}$

76. $\varnothing$
77. $\{0\}$
78. $\{2\}$

79. $\{5\}$
80. $\{1\}$
81. $\{17\}$

82. $\{-9\}$
83. $\{33\}$
84. $\{3\}$

85. $\{-4\}$

86. $\{3\}$

87. $\{-31\}$

88. $\{21\}$

89. $\{0\}$

90. $\{0\}$

9.6   Mixed Exercises

91. $\varnothing$

92. $\{64\}$

93. $\left\{\dfrac{3}{2}, \dfrac{5}{2}\right\}$

94. $\{7\}$

95. $\{0\}$

96. $\{8\}$

97. $\{-5\}$

98. $\{-3\}$

99. $\left\{1, \dfrac{3}{2}\right\}$

100. $\{4\}$

9.7   Complex Numbers

Objective 1

1. $7i$

2. $6i$

3. $-10i$

4. $i\sqrt{6}$

5. $i\sqrt{22}$

6. $5i\sqrt{2}$

7. $-3i\sqrt{7}$

8. $2i\sqrt{30}$

9. $3i\sqrt{2}$

10. $-3i\sqrt{3}$

11. $2i\sqrt{15}$

12. $15i\sqrt{2}$

13. $-5i\sqrt{5}$

14. $-6i\sqrt{2}$

15. $3i\sqrt{11}$

16. $6i\sqrt{30}$

17. $-25i$

18. $-9i\sqrt{2}$

19. -6

20. $-3\sqrt{5}$

21. $-7\sqrt{3}$

22. $i\sqrt{42}$

23. $i\sqrt{21}$

24. $i\sqrt{30}$

25. -6

26. $-i\sqrt{105}$

27. 5

28. $2i\sqrt{2}$

29. $2i$

30. 3

31. 3

32. $4i$

33. 5

34. $6i$

35. $-\sqrt{21}$

36. $\dfrac{i\sqrt{70}}{5}$

Objective 2

37. imaginary

38. imaginary

39. real

40. imaginary

41. real

42. imaginary

43. imaginary

44. real

45. imaginary

46. imaginary

Objective 3

47. $3 + 11i$

48. $8 + 6i$

49. $14 - 2i$

50. $7 + 6i$

51. 3

52. $-7 + 2i$

53. $-4 + i$

54. $5 + 3i$

55. $9 - 9i$

56. $-7 + 2i$

57. $-5 - i$

58. 1

59. $2 - 3i$

60. 10

61. $3\sqrt{3} - 4i\sqrt{2}$

62. $7 + 5i$

Objective 4

63. $11 + 13i$

64. $25 + 8i$

65. $44 + 12i$

66. 34

67. 53

68. $-13i$

69. $-8 + 6i$

70. $-1 - 2i\sqrt{6}$

71. $34 - 34i$

72. $7 + i$

Objective 5

73. $\dfrac{1}{5} + \dfrac{3}{5}i$

74. $\dfrac{37}{97} + \dfrac{38}{97}i$

75. $\dfrac{19}{17} + \dfrac{43}{17}i$

76. $-1-3i$

77. $\dfrac{10}{13}+\dfrac{11}{13}i$

78. $\dfrac{4}{5}-\dfrac{7}{5}i$

79. $\dfrac{15}{13}+\dfrac{16}{13}i$

80. $-\dfrac{3}{5}+\dfrac{6}{5}i$

81. $-\dfrac{1}{4}-\dfrac{1}{2}i$

82. $\dfrac{1}{5}+\dfrac{1}{5}i$

Objective 6

83. $-i$

84. i

85. 1

86. -1

87. $-i$

88. i

89. 1

90. -1

91. $-i$

92. 1

93. 1

94. $-i$

95. i

96. $-i$

97. $-i$

9.7   Mixed Exercises

98. $5i\sqrt{5}$

99. $2i$

100. i

101. i

102. $-7\sqrt{5}$

103. -8

104. $11-4i$

105. $13+i$

106. $-14+10i$

107. $2+4i$

108. $17+7i$

109. 29

110. $\dfrac{9}{13}+\dfrac{20}{13}i$

111. $\dfrac{47}{25}-\dfrac{21}{25}i$

Chapter 10

QUADRATIC EQUATIONS, INEQUALITIES, AND GRAPHS

10.1 Solving Quadratic Equations by the Square Root Property

Objective 1

1. $-7, 7$

2. $-11, 11$

3. $-30, 30$

4. $-9, 9$

5. $-2\sqrt{6}, 2\sqrt{6}$

6. no real number solution

7. $-3\sqrt{5}, 3\sqrt{5}$

8. $-5\sqrt{2}, 5\sqrt{2}$

9. $-6\sqrt{2}, 6\sqrt{2}$

10. $-7\sqrt{2}, 7\sqrt{2}$

11. no real number solution

12. $-5\sqrt{10}, 5\sqrt{10}$

13. $-\dfrac{3}{5}, \dfrac{3}{5}$

14. $-\dfrac{9}{8}, \dfrac{9}{8}$

15. $-\dfrac{3\sqrt{10}}{17}, \dfrac{3\sqrt{10}}{17}$

16. no real number solution

17. $-1.4, 1.4$

18. $-3.5, 3.5$

Objective 2

19. $-9, 1$

20. $12, -6$

21. $2, -6$

22. $2, 16$

23. 7

24. -8

25. $-5, -9$

26. $-3+3\sqrt{2}, -3-3\sqrt{2}$

27. $9+2\sqrt{7}, 9-2\sqrt{7}$

28. $2+3\sqrt{3}, 2-3\sqrt{3}$

29. no real number solution

30. $\dfrac{-5+4\sqrt{2}}{2}, \dfrac{-5-4\sqrt{2}}{2}$

31. $\dfrac{-2+3\sqrt{3}}{3}, \dfrac{-2-3\sqrt{3}}{3}$

32. $\dfrac{2+2\sqrt{7}}{3}, \dfrac{2-2\sqrt{7}}{3}$

33. $-\dfrac{7}{2}, -\dfrac{11}{2}$

34. no real number solution

35. $-72, 0$

36. $-25, 45$

Objective 3

37. $\{-i, i\}$

38. $\{-3i, 3i\}$

39. $\{-4i, 4i\}$

40. $\left\{\dfrac{7-3i}{6}, \dfrac{7+3i}{6}\right\}$

41. $\{-5i, 5i\}$

42. $\{-1-6i, -1+6i\}$

43. $\left\{-\dfrac{5}{2}i, \dfrac{5}{2}i\right\}$

44. $\{-11i, 11i\}$

45. $\{-9i, 9i\}$

46. $\left\{\dfrac{-16-7i}{8}, \dfrac{-16+7i}{8}\right\}$

10.1 Mixed Exercises

47. $\left\{-5+\sqrt{7},\, -5-\sqrt{7}\right\}$

48. $\{-15,\, 15\}$

49. $\left\{-3\sqrt{3},\, 3\sqrt{3}\right\}$

50. $\left\{\dfrac{-4+\sqrt{7}}{2},\, \dfrac{-4-\sqrt{7}}{2}\right\}$

51. $\{3+i,\, 3-i\}$

52. $\left\{\dfrac{-5+2i\sqrt{3}}{4},\, \dfrac{-5-2i\sqrt{3}}{4}\right\}$

53. $\left\{-\dfrac{10}{3},\, \dfrac{10}{3}\right\}$

54. $\left\{-\sqrt{10},\, \sqrt{10}\right\}$

55. $\{-3,\, 0\}$

56. $\left\{\dfrac{-7+3\sqrt{5}}{2},\, \dfrac{-7-3\sqrt{5}}{2}\right\}$

57. $\left\{\dfrac{-7+3\sqrt{17}}{4},\, \dfrac{-7-3\sqrt{17}}{4}\right\}$

58. $\left\{1+i\sqrt{2},\, 1-i\sqrt{2}\right\}$

59. $\{1,\, 4\}$

60. $\left\{-\dfrac{3}{2},\, 2\right\}$

10.2 Solving Quadratic Equations by Completing the Square

Objective 1

1. $-3, 5$

2. $-2, 7$

3. $0, -6$

4. $-4, 1$

5. no real number solution

6. $-4 + 2\sqrt{3}, -4 - 2\sqrt{3}$

7. $3, 7$

8. $2 + \sqrt{6}, 2 - \sqrt{6}$

9. $-2 + \sqrt{6}, -2 - \sqrt{6}$

10. $-1, 4$

11. $\dfrac{-5 + 3\sqrt{5}}{2}, \dfrac{-5 - 3\sqrt{5}}{2}$

12. $-9, 7$

13. 4

14. $-11, 1$

15. $\dfrac{1}{2}, -\dfrac{7}{2}$

16. $\dfrac{1 + \sqrt{11}}{2}, \dfrac{1 - \sqrt{11}}{2}$

Objective 2

17. $-2, 0$

18. $-2, 7$

19. $\dfrac{5}{2}, 4$

20. $\dfrac{3 + \sqrt{15}}{3}, \dfrac{3 - \sqrt{15}}{3}$

21. $\dfrac{-3 + \sqrt{57}}{6}, \dfrac{-3 - \sqrt{15}}{6}$

22. $\dfrac{-3 + \sqrt{11}}{2}, \dfrac{-3 - \sqrt{11}}{2}$

23. no real number solution

24. $-\dfrac{3}{2}, \dfrac{5}{3}$

25. $\dfrac{1}{2}, -7$

26. $-2, -\dfrac{1}{3}$

27. $\dfrac{3 + \sqrt{13}}{2}, \dfrac{3 - \sqrt{13}}{2}$

28. $-4, 2$

29. $\dfrac{3 + \sqrt{15}}{3}, \dfrac{3 - \sqrt{15}}{3}$

30. $-1, \dfrac{2}{3}$

31. no real number solution

32. $\dfrac{-2 + \sqrt{10}}{6}, \dfrac{-2 - \sqrt{10}}{6}$

Objective 3

33. $\dfrac{3 + \sqrt{69}}{6}, \dfrac{3 - \sqrt{69}}{6}$

34. 2

35. $-\dfrac{3}{2}, \dfrac{1}{2}$

36. no real number solution

37. $\dfrac{-1 + \sqrt{65}}{2}, \dfrac{-1 - \sqrt{65}}{2}$

38. $-8, 2$

39. $-8, -2$

40. $-2 + \sqrt{2}, -2 - \sqrt{2}$

41. $-3, 2$

42. $\dfrac{-1 + 3\sqrt{5}}{2}, \dfrac{-1 - 3\sqrt{5}}{2}$

10.2 Mixed Exercises

43. $0, -3$

44. $-2, 5$

45. $-1, -2$

46. no real number solution

47. $-2, 3$

48. no real number solution

49. $\dfrac{1+\sqrt{3}}{2}, \dfrac{1-\sqrt{3}}{2}$

50. $3, 1$

51. $3+\sqrt{3}, 3-\sqrt{3}$

52. $5+\sqrt{17}, 5-\sqrt{17}$

53. $-2+\sqrt{29}, -2-\sqrt{29}$

54. $\dfrac{2+\sqrt{13}}{3}, \dfrac{2-\sqrt{13}}{3}$

10.3 Solving Quadratic Equations by the Quadratic Formula

Objective 2

1. $\{1, 6\}$

2. $\{3, 9\}$

3. $\{-7, 2\}$

4. $\{-5, 8\}$

5. $\{4\}$

6. $\left\{\dfrac{3}{5}, 2\right\}$

7. $\left\{\dfrac{4}{3}, \dfrac{3}{2}\right\}$

8. $\{-4, 2\}$

9. $\left\{-\dfrac{3}{4}, \dfrac{3}{4}\right\}$

10. $\{2, 4\}$

11. $\left\{\dfrac{-5+3\sqrt{5}}{10}, \dfrac{-5-3\sqrt{5}}{10}\right\}$

12. $\left\{\dfrac{-1+\sqrt{2}}{2}, \dfrac{-1-\sqrt{2}}{2}\right\}$

13. $\left\{\dfrac{3+\sqrt{6}}{3}, \dfrac{3-\sqrt{6}}{3}\right\}$

14. $\left\{\dfrac{2+i\sqrt{2}}{2}, \dfrac{2-i\sqrt{2}}{2}\right\}$

15. $\left\{-\dfrac{1}{4}, 3\right\}$

16. $\left\{\dfrac{2}{3}, \dfrac{3}{2}\right\}$

Objective 3

17. D

18. D

19. D

20. A

21. D

22. C

23. B

24. D

25. B

26. C

27. C

28. A

29. C

30. A

10.3 Mixed Exercises

31. $\left\{1+i\sqrt{2}, 1-i\sqrt{2}\right\}$

32. $\left\{\dfrac{1+\sqrt{13}}{2}, \dfrac{1-\sqrt{13}}{2}\right\}$

33. $\{-2, 14\}$

34. $\{-2, 1\}$

35. A

36. C

37. A

38. A

39. B

40. D

10.4 Equations Quadratic in Form

Objective 1

1. $\{-3, 2\}$

2. $\left\{-\dfrac{5}{3}, 3\right\}$

3. $\left\{-\dfrac{7}{5}, 2\right\}$

4. $\left\{-\dfrac{7}{2}, -\dfrac{1}{3}\right\}$

5. $\left\{-3, -\dfrac{3}{2}\right\}$

6. $\left\{-7, \dfrac{5}{4}\right\}$

7. $\left\{-\dfrac{7}{2}, 4\right\}$

8. $\left\{\dfrac{2}{3}\right\}$

9. $\{-3, 5\}$

10. $\left\{-3, -\dfrac{1}{5}\right\}$

11. $\left\{\dfrac{9}{7}, 4\right\}$

12. $\left\{-\dfrac{35}{4}, -3\right\}$

13. $\left\{-7, \dfrac{5}{2}\right\}$

14. $\left\{-\dfrac{5}{2}, 1\right\}$

15. $\left\{\dfrac{8}{3}, 6\right\}$

16. $\{-1, 2\}$

Objective 2

17. 4 in. by 8 in.

18. 5 mph

19. 2 mph

20. 10 hr, 15 hr

21. 50 mph

22. 15 hr, 30 hr

23. 550 mph

24. 3 or $-\dfrac{40}{11}$

25. bike: 12 mph; hike: 2 mph

26. 5 or $\dfrac{1}{5}$

Objective 3

27. $\{2\}$

28. $\{2, 5\}$

29. $\left\{\dfrac{3}{2}\right\}$

30. $\{3, 5\}$

31. $\left\{\dfrac{3}{2}\right\}$

32. $\left\{\dfrac{1}{4}\right\}$

33. $\left\{\dfrac{7}{3}, 7\right\}$

34. $\left\{\dfrac{1}{4}, \dfrac{1}{3}\right\}$

35. $\left\{\dfrac{1}{16}, \dfrac{1}{9}\right\}$

36. $\{4\}$

Objective 4

37. $\{-4, -3, 3, 4\}$

38. $\left\{-1, -\dfrac{3}{4}, \dfrac{3}{4}, 1\right\}$

39. $\{-6, 14\}$

40. $\{-3, -2, 2, 3\}$

41. $\{-9, -7\}$

42. $\{0, 9\}$

43. $\{9\}$

44. $\left\{-\dfrac{1}{3}, \dfrac{1}{4}\right\}$

45. $\{-2, 2\}$

46. $\left\{-3, -\sqrt{7}, \sqrt{7}, 3\right\}$

47. $\left\{-\sqrt{5}, -\dfrac{1}{2}, \dfrac{1}{2}, \sqrt{5}\right\}$

48. $\left\{-\dfrac{17}{3}, -4\right\}$

49. $\left\{-3\sqrt{2}, -\sqrt{2}, \sqrt{2}, 3\sqrt{2}\right\}$

50. $\left\{-2\sqrt{5}, 2\sqrt{5}\right\}$

51. $\left\{-2\sqrt{2}, 2\sqrt{2}, -\sqrt{3}, \sqrt{3}\right\}$

52. $\left\{-\sqrt{10}, -\sqrt{2}, \sqrt{2}, \sqrt{10}\right\}$

53. $\left\{-\sqrt{6}, -2, 2, \sqrt{6}\right\}$

54. $\left\{-3, -\sqrt{3}, \sqrt{3}, 3\right\}$

55. $\{-1, 1\}$

56. $\left\{-\sqrt{3}, 0, \sqrt{3}\right\}$

57. $\left\{-\sqrt{3}, 0, \sqrt{3}\right\}$

58. $\left\{-10, -\dfrac{1}{6}, \dfrac{1}{6}, 10\right\}$

10.4 Mixed Exercises

59. 18.6 hr

60. 25.2 hr, 37.2 hr

61. 5.9 mph

62. $\left\{-\dfrac{7}{2}, \dfrac{1}{5}\right\}$

63. $\{3, 4\}$

64. $\left\{-7, -\dfrac{7}{2}\right\}$

65. $\{-3, 2\}$

66. $\left\{-\dfrac{4}{3}, 2\right\}$

67. $\{-2, -1, 4, 5\}$

68. $\{9\}$

69. $\left\{-2, -\dfrac{\sqrt{3}}{2}, \dfrac{\sqrt{3}}{2}, 2\right\}$

10.5 Formulas and Further Applications

Objective 1

1. $k = \dfrac{D^2}{h}$

2. $d = \dfrac{k^2 l^2}{F^2}$

3. $k = \dfrac{p^2 g}{l}$

4. $z = \dfrac{p\sqrt{6}}{y}$

5. $p = \dfrac{900a}{s^2}$

6. $c = \dfrac{(a-1)^2}{b}$

7. $t = \dfrac{\pm\sqrt{2gy}}{g}$

8. $t = \dfrac{\pm\sqrt{mxF}}{F}$

9. $x = \dfrac{\pm\sqrt{2Fk}}{k}$

Objective 2

10. 12 ft

11. 10 ft

12. south: 72 mi; east: 54 mi

13. north: 57 mi; west: 76 mi

14. 36 ft

15. 60 ft

16. 34 cm

17. 8 ft

18. 12 m

19. 25 ft

20. 48 cm

21. 24 in.

Objective 3

22. 3 m by 5 m

23. 12 in. by 16 in.

24. 3 cm

25. 7 in. by 11 in.

26. 12 in. by 15 in.

27. 8 in. by 5 in. by 4 in.

28. 9 in. by 14 in.

29. 16 cm by 10 cm

30. 2 ft

31. 3 ft

32. 2.5 in.

33. 2.5 ft

Objective 4

34. 17.1 hr

35. 1.4 min

36. 9.1 sec

37. 1.2 sec

10.5 Mixed Exercises

38. $t = \dfrac{p \pm \sqrt{p^2 + 4pq}}{2}$

39. $m = \dfrac{-x \pm \sqrt{x^2 + 4xy}}{2}$

40. $q = \dfrac{-k \pm k\sqrt{5}}{2p}$

41. $a = \dfrac{-c \pm c\sqrt{2}}{b}$

42. 4 ft

43. 15 cm

44. 13 in.

45. 1.5 ft

46. 27 items

10.6 Graphs of Quadratic Functions

Objectives 1 and 2

1. $(0, -2)$

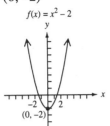

$f(x) = x^2 - 2$

5. $(0, 2)$

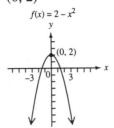

$f(x) = 2 - x^2$

9. $(-3, -1)$

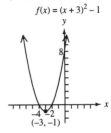

$f(x) = (x + 3)^2 - 1$

2. $(0, 2)$

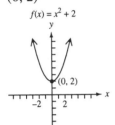

$f(x) = x^2 + 2$

6. $(0, 5)$

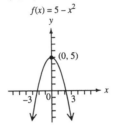

$f(x) = 5 - x^2$

10. $(2, 1)$

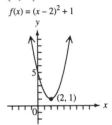

$f(x) = (x - 2)^2 + 1$

3. $(0, 3)$

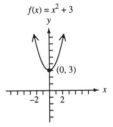

$f(x) = x^2 + 3$

7. $(-2, 0)$

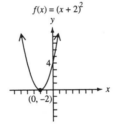

$f(x) = (x + 2)^2$

4. $(0, -4)$

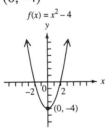

$f(x) = x^2 - 4$

8. $(3, 0)$

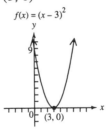

$f(x) = (x - 3)^2$

Objective 3

11. up; wider

12. up; narrower

13. down; narrower

14. down; narrower

15. up; wider

16. up; wider

17. down; narrower

18. down; wider

19. up; narrower

20. down; same

Objective 4

21. quadratic; negative

22. linear; positive

23. quadratic; positive

10.6 Mixed Exercises

24. $(1, 0)$

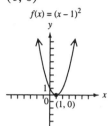

$f(x) = (x - 1)^2$

25. $(-2, 3)$

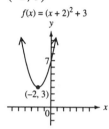

$f(x) = (x + 2)^2 + 3$

26. $(3, -1)$

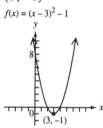

$f(x) = (x - 3)^2 - 1$

27. $(-3, 0)$

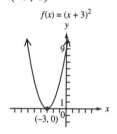

$f(x) = (x + 3)^2$

28. up; narrower; $(0, 0)$

29. down, same; $(0, 2)$

30. up; wider; $(1, 0)$

31. up; narrower; $(-1, -2)$

32. down; wider; $(-3, 0)$

33. up; narrower; $(3, 1)$

10.7 More About Parabolas; Applications

Objectives 1 and 2

1. $(-3, 1)$

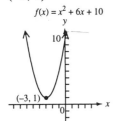

2. $(3, -5)$

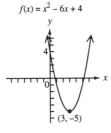

3. $(4, 6)$

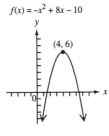

4. $\left(\dfrac{3}{2}, -\dfrac{1}{4}\right)$

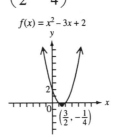

5. $(-1, -1)$

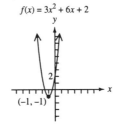

6. $(1, 3)$

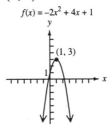

7. $(-2, 1)$

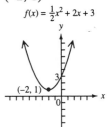

8. $(-2, -2)$

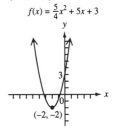

Objective 3

9. 0

10. 0

11. 0

12. 2

13. 0

14. 2

15. 2

16. 0

17. 1

18. 2

Objective 4

19. 25 units; $3650

20. 25 units; $1000

21. 50 pots; $200

22. 64 ft; 2 sec

23. 16 ft; 1 sec

24. 286 ft; $\dfrac{3}{2}$ sec

25. 256 ft; $\dfrac{5}{2}$ sec

26. 24 (a square)

27. 32 (a square)

28. 6 (a square)

Objective 5

29.   domain: $[0, \infty)$
range: $(-\infty, \infty)$
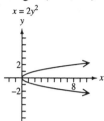

32.   domain: $[-3, \infty)$
range: $(-\infty, \infty)$
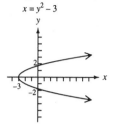

35.   domain: $[3, \infty)$
range: $(-\infty, \infty)$
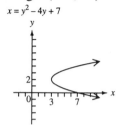

30.   domain: $(-\infty, 0]$
range: $(-\infty, \infty)$
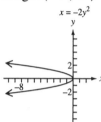

33.   domain: $[0, \infty)$
range: $(-\infty, \infty)$
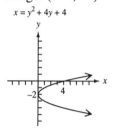

36.   domain: $[-1, \infty)$
range: $(-\infty, \infty)$
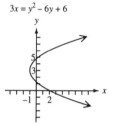

31.   domain: $(-\infty, 2]$
range: $(-\infty, \infty)$
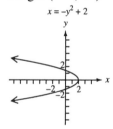

34.   domain: $(-\infty, 0]$
range: $(-\infty, \infty)$
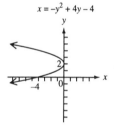

10.7 Mixed Exercises

37. $(-3, -1)$
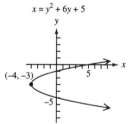

38. $(-4, -3)$
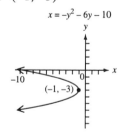

39. $(-1, -3)$

40. $\left(-1, -\dfrac{5}{2}\right)$

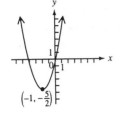

$f(x) = 2x^2 + 4x - \frac{1}{2}$

$\left(-1, -\frac{5}{2}\right)$

41. 2

42. 0

43. 8 units; $98

44. 4 (square)

45. 46 and 46

10.8 Quadratic and Rational Inequalities

Objective 1

1. $(-\infty, -3] \cup [2, \infty)$

2. $(-3, 2)$

3. $(-2, 5)$

4. $(-\infty, -3] \cup [2, \infty)$

5. $(-\infty, -4) \cup (-3, \infty)$

6. $[-1, 2]$

7. $\left(-1, \dfrac{3}{2}\right)$

8. $\left(-\infty, -\dfrac{3}{2}\right) \cup (4, \infty)$

9. $\left(-\infty, -\dfrac{2}{3}\right) \cup \left(-\dfrac{1}{2}, \infty\right)$

10. $\left(-\infty, -\dfrac{3}{2}\right) \cup \left(\dfrac{1}{4}, \infty\right)$

11. $(-\infty, \infty)$

12.

13.

14. $(-\infty, \infty)$

15.

16.

Objective 2

17. $(-\infty, -4] \cup [-1, 2]$

18. $(-\infty, -2) \cup (1, 2)$

19. $(-\infty, -5] \cup [-3, 1]$

20. $[-2, 1] \cup [3, \infty)$

21. $(-\infty, -3] \cup [-1, 4]$

22. $(2, 4) \cup (6, \infty)$

23. $\left(-\infty, -\dfrac{3}{2}\right] \cup \left[-\dfrac{1}{3}, \dfrac{1}{2}\right]$

24. $\left(-\dfrac{1}{4}, \dfrac{1}{6}\right) \cup \left(\dfrac{7}{3}, \infty\right)$

25. $\left[\dfrac{3}{4}, \dfrac{10}{3}\right] \cup \left[\dfrac{7}{2}, \infty\right)$

26. $(-\infty, -1) \cup \left(1, \dfrac{7}{3}\right)$

Objective 3

27. $(-\infty, 1) \cup [8, \infty)$

28. $(-\infty, -3) \cup [-1, \infty)$

29. $(-1, 4]$

30. $\left(-2, -\dfrac{2}{3}\right)$

31. $\left(-\dfrac{5}{3}, -1\right)$

32. $\left(-\infty, \dfrac{5}{2}\right] \cup (3, \infty)$

33. $(-\infty, -2)$

34. $(-4, \infty)$

35. $\left(\dfrac{2}{3}, \dfrac{8}{3}\right]$

36. $\left(-\infty, -\dfrac{19}{9}\right] \cup \left(-\dfrac{5}{3}, \infty\right)$

37. $\left(-\infty, \dfrac{1}{4}\right] \cup \left(\dfrac{3}{2}, \infty\right)$

38. $\left[0, \dfrac{3}{4}\right)$

39. $\left[-\dfrac{3}{2}, -1\right)$

40. $\left(2, \dfrac{8}{3}\right]$

41. $(-\infty, -2] \cup \left(-\dfrac{1}{3}, \infty\right)$

42. $(-\infty, 3) \cup [8, \infty)$

43. $[-2, 3)$

44. $\left(\dfrac{15}{13}, \dfrac{5}{3}\right)$

10.8 Mixed Exercises

45. $\left[\dfrac{1}{3}, \dfrac{2}{5}\right]$

46. $\left(-\infty, -\dfrac{2}{3}\right] \cup \left[\dfrac{5}{2}, \infty\right)$

47. $\left[-\dfrac{3}{2}, \dfrac{3}{2}\right]$

48. $\left(-\infty, -\dfrac{3}{4}\right] \cup \left[\dfrac{3}{4}, \infty\right)$

49. $\left(-\infty, -\dfrac{1}{2}\right) \cup \left(\dfrac{1}{4}, \infty\right)$

50. $\left(-3, \dfrac{1}{6}\right)$

51. $[-3, -1] \cup [2, \infty)$

52. $(-\infty, -2) \cup (2, 3)$

53. $(-\infty, -1) \cup \left(\dfrac{1}{2}, 3\right)$

54. $\left[\dfrac{3}{4}, 1\right] \cup [3, \infty)$

55. $\left(-\dfrac{3}{2}, \dfrac{1}{3}\right) \cup (2, \infty)$

56. $\left(-\infty, -\dfrac{5}{2}\right] \cup \left[-1, \dfrac{4}{5}\right]$

57. $\left[-\dfrac{3}{2}, 0\right)$

58. $(-\infty, 0) \cup \left[\dfrac{1}{2}, \infty\right)$

59. $\left(0, \dfrac{3}{5}\right]$

60. $(-\infty, 0) \cup \left[\dfrac{3}{8}, \infty\right)$

61.

62. $(-\infty, \infty)$

Chapter 11

EXPONENTIAL AND LOGARITHMIC FUNCTIONS

11.1 Inverse Functions

Objective 1

1. not one-to-one

2. not one-to-one

3. $\{(-1, 2), (1, -2), (3, 1), (-3, -1)\}$

4. $\{(-3, 6), (-2, 4), (-1, 2), (0, 0)\}$

5. not one-to-one

6. not one-to-one

7. $\{(1, -1), (2, -2), (3, -3)\}$

8. $\{(1, -3), (2, -2), (3, -1), (4, 0)\}$

9. $\{(2, 3), (-2, -3), (3, 2), (-3, -2)\}$

10. not one-to-one

Objective 2

11. one-to-one

12. not one-to-one

13. not one-to-one

14. not one-to-one

15. one-to-one

16. one-to-one

17. not one-to-one

18. one-to-one

Objective 3

19. $f^{-1}(x) = \dfrac{x+5}{2}$

20. $f^{-1}(x) = \dfrac{x+5}{3}$

21. not one-to-one

22. not one-to-one

23. $f^{-1}(x) = x^2 + 1;\ x \geq 0$

24. $f^{-1}(x) = \dfrac{x^2}{12};\ x \geq 0$

25. $f^{-1}(x) = \sqrt[3]{x+1}$

26. $f^{-1}(x) = \dfrac{\sqrt[3]{4x+12}}{2}$

27. not one-to-one

28. $f^{-1}(x) = \dfrac{3+x}{x}$

Objective 4

29.

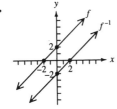

30.

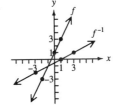

31. not one-to-one

32.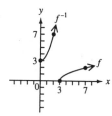

33. not one-to-one

11.1   Mixed Exercises

37. $\{(5, 3), (9, 2), (7, 4)\}$

38. $\{(0, 0), (1, 1), (-1, -1), (2, 2), (-2, -2)\}$

39. not one-to-one

40. $\{(-1, -3), (0, -2), (1, -1), (2, 0)\}$

41. $f^{-1}(x) = \dfrac{4 - x}{2}$

42. $f^{-1}(x) = x^2 - 2;\ x \ge 0$

43. not one-to-one

44. $f^{-1}(x) = \sqrt[3]{x + 5}$

34. not one-to-one

35.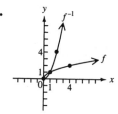

36. not one-to-one

45.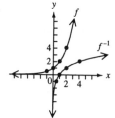

46.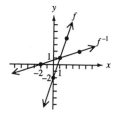

11.2 Exponential Functions

Objective 1

1. exponential function

2. not an exponential function

3. not an exponential function

4. exponential function

5. not an exponential function

6. not an exponential function

7. exponential function

8. exponential function

9. not an exponential function

10. not an exponential function

Objective 2

11.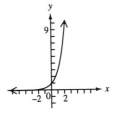

13.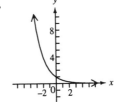

15.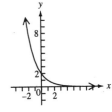

12.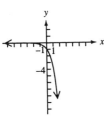

14.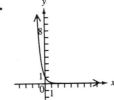

16.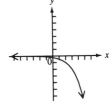

Objective 3

17. $\left\{\dfrac{3}{2}\right\}$

18. $\left\{\dfrac{2}{3}\right\}$

19. $\{2\}$

20. $\left\{\dfrac{3}{4}\right\}$

21. $\left\{\dfrac{1}{2}\right\}$

22. $\{1\}$

23. $\left\{\dfrac{1}{2}\right\}$

24. $\left\{\dfrac{3}{4}\right\}$

25. $\left\{\dfrac{1}{2}\right\}$

26. $\left\{-\dfrac{1}{2}\right\}$

Objective 4

27. 515 geese

28. 7.5 in.

29. 3650 bacteria

30. 35,000

31. 1 g

32. 200 bacteria

33. 1000 bacteria

34. 256,000

11.2   Mixed Exercises

35. $\left\{ \dfrac{3}{2} \right\}$

36. $\left\{ -\dfrac{5}{4} \right\}$

37. $\left\{ \dfrac{1}{4} \right\}$

38. $\{1\}$

39. $\{-3\}$

40. $\{-2\}$

41.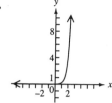

42.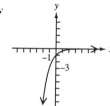

43. $27,048.14

44. 250,000 barrels

11.3 Logarithmic Functions

Objective 1

1. 3

2. 2

3. −1

4. $\dfrac{1}{2}$

5. −2

6. −4

7. $\dfrac{1}{2}$

8. $\dfrac{1}{2}$

9. −2

10. $\dfrac{3}{4}$

Objective 2

11. $\log_3 9 = 2$

12. $\log_5 \sqrt[3]{5} = \dfrac{1}{3}$

13. $4^{-2} = \dfrac{1}{16}$

14. $16^{1/4} = 2$

15. $\log_{10} \dfrac{1}{100} = -2$

16. $5^2 = 25$

17. $9^{1/2} = 3$

18. $\log_9 3 = \dfrac{1}{2}$

19. $10^{-3} = .001$

20. $\log_2 \dfrac{1}{128} = -7$

Objective 3

21. $\{6\}$

22. $\left\{\dfrac{1}{5}\right\}$

23. $\{5\}$

24. $\{3\}$

25. $\left\{\dfrac{1}{32}\right\}$

26. $\{2\}$

27. $\{10\}$

28. $\{4\}$

29. $\{1\}$

30. $\{1\}$

Objective 4

31.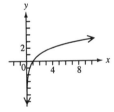

33.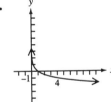

35.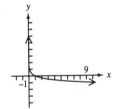

37.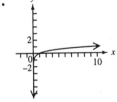

32.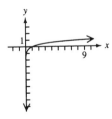

34.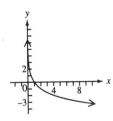

36.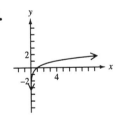

38.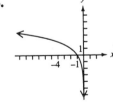

Objective 5

39. 20 squirrels

40. $600

41. 1200 cm/sec

42. 110 decibels

43. 8 students

44. 40 fish

45. 12 squirrels

46. 200 applicants

47. 335,000 items

48. 100 mites

11.3   Mixed Exercises

49. $\log_{1/2} 8 = -3$

50. $5^{-4} = .0016$

51. $\left\{ \dfrac{3}{5} \right\}$

52. $\{81\}$

53. $\{0\}$

54. $\{16\}$

55.

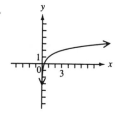

56. 239 foxes

57. 16 fish

58. $100,000

11.4 Properties of Logarithms

Objective 1

1. $\log_3 6 + \log_3 5$

2. $\log_2 5 + \log_2 3$

3. $\log_7 5 + \log_7 m$

4. $\log_2 6 + \log_2 x + \log_2 y$

5. $1 + \log_6 r$

6. $\log_3 2 + \log_3 p$

7. $\log_4 21$

8. $\log 12$

9. $\log_7 66 y^3$

10. $\log_7 120 r^5$

Objective 2

11. $\log_2 7 - \log_2 9$

12. $\log_4 5 - \log_4 8$

13. $\log_3 m - \log_3 n$

14. $\log p - \log r$

15. $\log_6 k - \log_6 3$

16. $\log_3 10 - \log_3 x$

17. $3 - \log_2 m$

18. $1 - \log_5 x$

19. $\log_2 \dfrac{7q^2}{5}$

20. $\log 3x$

21. $\log_7 \dfrac{3}{5r^4}$

22. $\log_9 \dfrac{2}{y^2}$

Objective 3

23. $2\log_5 3$

24. $3\log_3 4$

25. $3\log_2 5$

26. $7\log_m 2$

27. $\dfrac{1}{2}\log_b 5$

28. $\dfrac{1}{3}\log_3 7$

29. $\dfrac{1}{2}$

30. $\dfrac{1}{3}$

31. 1

32. 1

Objective 4

33. $2 + 3\log_2 p$

34. $2 + 3\log_3 x$

35. $\dfrac{1}{3}(\log_a 2 + \log_a k)$

36. $\log_b 2 + \log_b r - \log_b (r-1)$

37. $2 - \log_2 3$

38. $\log_3 5 - 2$

39. $\log_5 7 + 3\log_5 m - \log_5 8 - \log_5 y$

40. $\log_7 8 + 7\log_7 r - \log_7 3 - 3\log_7 a$

41. $\log 14x^2$

42. $\log_a 8r^3$

43. $\log_b \dfrac{3q}{2p}$

44. $\log \dfrac{4k}{3j}$

45. $\log_4 \dfrac{5}{y}$

46. $\log_6 \dfrac{7}{m}$

47. 0

48. 0

11.4 Mixed Exercises

49. $3 + \log_2 p$

50. $\dfrac{3}{2}$

51. $1 - \log_4 9$

52. $4 \log_5 k$

53. 2

54. $\dfrac{1}{2} \log_5 3 + \dfrac{1}{2} \log_5 p$

55. $\log_7 3 - 1$

56. $\log_4 3 + \log_4 m - \log_4 (m + 2)$

57. 1

58. $\log_3 10q^4$

59. $\log_5 16 y^2$

60. 1

61. $\log_5 \dfrac{1}{m^4}$ or $-\log_5 m^4$

62. $\log_2 \dfrac{1}{y}$ or $-\log_2 y$

11.5 Common and Natural Logarithms

Objective 1

1. 1.7576
2. .9031
3. 2.9262
4. −1.0390
5. 5.4472
6. 2.9025
7. −4.4970
8. 1.7853
9. −3.0186
10. 4.9401
11. 1.3424
12. −.4558
13. 2.8848
14. 3.7395
15. −3.0814
16. −3
17. 6.0133
18. .6022

Objective 2

19. 5.6
20. 7.3
21. 6.7
22. 8.8
23. 4.2
24. 10.1
25. 1.3×10^{-3}
26. 4.0×10^{-4}
27. 6.3×10^{-6}
28. 5.0×10^{-2}
29. 3.2×10^{-7}
30. 6.3×10^{-11}

Objective 3

31. −2.1203
32. 4.6052
33. 1.7918
34. 6.0591
35. −4.3428
36. 4.2341
37. 1.3863
38. −2.2828
39. 6.7731
40. 4.3347
41. −4.6052
42. −6.1469

Objective 4

43. 600
44. 618
45. 663
46. 733
47. 100 g
48. 74.1 g
49. 40.7 g
50. 16.5 g
51. 10,000
52. 11,600
53. 12,200
54. 33,200

11.5   Mixed Exercises

55. −1.0286
56. 3.9120
57. −7.1234
58. 4.7795
59. 8.4
60. 1.6×10^{-5}
61. 3814 termites
62. 20,000

11.6 Exponential and Logarithmic Equations; Further Applications

Objective 1

1. {.488} **3.** {1.232} **5.** {4.292} **7.** {3.936} **9.** {−.183}

2. {.517} **4.** {−1.395} **6.** {3.472} **8.** {−1.339} **10.** {1.380}

Objective 2

11. {5}

12. {3}

13. $\left\{\dfrac{1}{4}\right\}$

14. {3}

15. $\left\{-\dfrac{5}{6}\right\}$

16. {2}

17. $\left\{\sqrt[4]{10}\right\}$

18. {625}

19. $\left\{\dfrac{9}{8}\right\}$

20. {1}

Objective 3

21. $1259.71 **23.** $12,905.41 **25.** $5540.08

22. $40,262.75 **24.** $6391.88 **26.** $111,295.13

Objective 4

27. 8.25 g **28.** 29 yr **29.** 264 g **30.** 22 yr

Objective 5

31. .6131 **34.** 3.2266 **37.** 3.8074 **40.** −1.7297 **43.** .4150

32. 3.3219 **35.** 4.5110 **38.** 3.4130 **41.** −1.2770 **44.** 1.6309

33. 1.7124 **36.** 1.1887 **39.** −2.5850 **42.** −3.9694

11.6   Mixed Exercises

45. {1.7227} **48.** {4.1885} **51.** $114,237.51 **54.** 7 yr

46. {−.6309} **49.** {1000} **52.** $68,098.05 **55.** 1.1950

47. {3.3802} **50.** {1} **53.** 18 yr **56.** 1.5

Chapter 12

NONLINEAR FUNCTIONS, CONIC SECTIONS, AND NONLINEAR SYSTEMS

12.1 Additional Graphs of Functions; Composition

Objective 1

1. $f(x) = |x - 2| + 3$
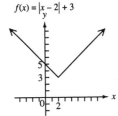

4. $f(x) = -|x + 3| - 2$
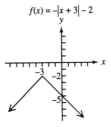

7. $f(x) = |x - 3| - 2$
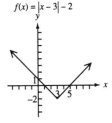

2. $f(x) = \sqrt{x + 3}$
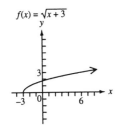

5. $f(x) = \sqrt{5 - x}$
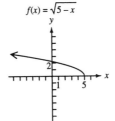

8. $f(x) = \sqrt{x} + 3$
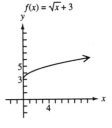

3. $f(x) = \frac{1}{x - 1}$
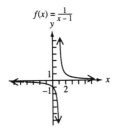

6. $f(x) = \frac{1}{x} + 3$
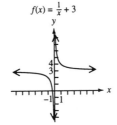

Objective 3

9. $3x^2 + 1$

10. $x^2 - 1$

11. $2x^4 + 2x^2$

12. $\dfrac{2x^2}{x^2 + 1}$

Objective 4

13. 12

14. 28

15. 38

16. 23

17. 4

18. $9x^2 + 12x + 7$

19. $3x + 14$

20. $x^2 + 8x + 19$

21. $3x^2 + 11$

22. $3x + 6$

12.1   Mixed Exercises

23. $f(x) = |x - 4| + 2$

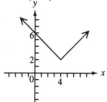

24. $f(x) = -\sqrt{x - 3} - 3$

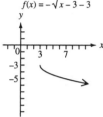

25. $f(x) = \dfrac{1}{x - 3}$

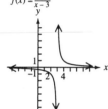

26. $f(x) = -|x - 3| + 3$

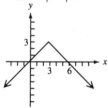

27. $f(x) = \sqrt{4 - x^2}$

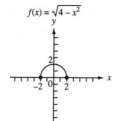

28. $f(x) = -\sqrt{1 - x^2}$

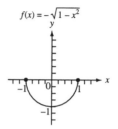

29. 7

30. 10

31. $x^2 + 3$

32. $x^2 - 6x + 15$

33. -11

34. -10

12.2 The Circle and the Ellipse

Objective 1

1. $(x+3)^2 + (y-2)^2 = 25$

2. $(x-1)^2 + (y-4)^2 = 4$

3. $x^2 + (y-5)^2 = 9$

4. $(x-6)^2 + (y-2)^2 = 9$

5. $(x+5)^2 + (y-4)^2 = 16$

6. $(x-7)^2 + (y-1)^2 = 4$

7. $(x-3)^2 + (y+4)^2 = 25$

8. $(x-2)^2 + (y-2)^2 = 36$

9. $(x-1)^2 + (y-3)^2 = 25$

10. $(x+2)^2 + (y+2)^2 = 9$

Objective 2

11. $(2, -4); 3$

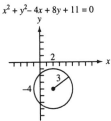

12. $(-3, 2); 1$

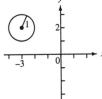

13. $(3, -5); 8$

14. $(2, 1); 6$

15. $(-2, -3); 4$

16. $(5, -6) ; 3$

17. $(4, 1); \sqrt{2}$

18. $(2, -4); 3$

19. $(2, -1); \sqrt{7}$

20. $(-5, -2); 6$

Objectives 3 and 4

21.

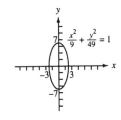

22.

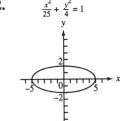

23. $\frac{x^2}{25} + \frac{y^2}{36} = 1$

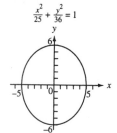

24. $\frac{x^2}{4} + \frac{y^2}{9} = 1$

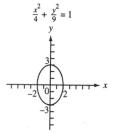

25. $\frac{x^2}{16} + \frac{y^2}{25} = 1$

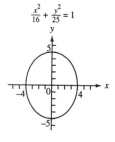

26. $\frac{x^2}{36} + \frac{y^2}{9} = 1$

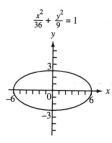

27. $\frac{x^2}{25} + \frac{y^2}{64} = 1$

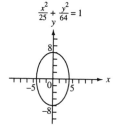

28. $\frac{x^2}{4} + \frac{y^2}{16} = 1$

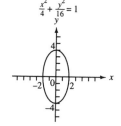

12.2   Mixed Exercises

29. $(x+2)^2 + (y+4)^2 = 25$

30. $x^2 + (y-3)^2 = 2$

31. $(-4, -2); 7$

32. $(3, -2); \dfrac{3}{2}$

33. $\frac{x^2}{16} + \frac{y^2}{49} = 1$

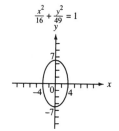

34. $\frac{x^2}{25} + \frac{y^2}{81} = 1$

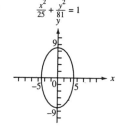

12.3   The Hyperbola and Other Functions Defined by Radicals

Objectives 1 and 2

1. $\frac{x^2}{9} - \frac{y^2}{16} = 1$

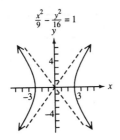

4. $\frac{y^2}{25} - \frac{x^2}{16} = 1$

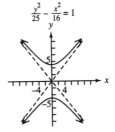

7. $\frac{x^2}{25} - \frac{y^2}{4} = 1$

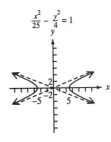

2. $\frac{x^2}{25} - \frac{y^2}{9} = 1$

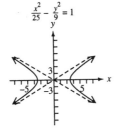

5. $\frac{x^2}{36} - \frac{y^2}{49} = 1$

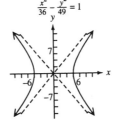

8. $\frac{x^2}{25} - \frac{y^2}{81} = 1$

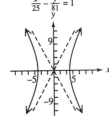

3. $\frac{y^2}{4} - \frac{x^2}{9} = 1$

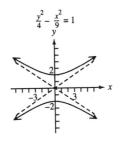

6. $\frac{y^2}{4} - \frac{x^2}{4} = 1$

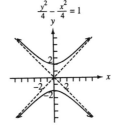

Objective 3

9. hyperbola

10. ellipse

11. parabola

12. hyperbola

13. hyperbola

14. parabola

15. circle

16. circle

17. circle

18. ellipse

Objective 4

19. $f(x) = \sqrt{36 - x^2}$

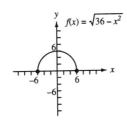

20. $f(x) = \sqrt{25 - x^2}$

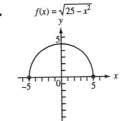

21. $f(x) = -\sqrt{4 - x^2}$

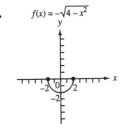

22. $f(x) = -\sqrt{9 - x^2}$

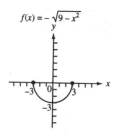

24. $f(x) = -3\sqrt{1 + x^2/25}$

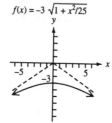

26. $f(x) = -5\sqrt{1 - x^2/9}$

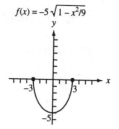

23. $f(x) = \sqrt{1 + x^2/4}$

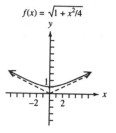

25. $f(x) = \sqrt{9 - 9x^2}$

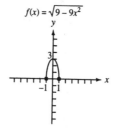

12.3   Mixed Exercises

27. parabola

28. circle

29. circle

30. parabola

31. hyperbola

32. parabola

12.4 Nonlinear Systems of Equations

Objective 1

1. $\left\{(4, -1), \left(\dfrac{16}{5}, -\dfrac{13}{5}\right)\right\}$

2. $\{(12, -17), (2, 3)\}$

3. $\left\{\left(\dfrac{1}{4}, \dfrac{3}{2}\right), (-1, 1)\right\}$

4. $\left\{(-2, 1), \left(-\dfrac{50}{17}, -\dfrac{31}{17}\right)\right\}$

5. $\{(5, 2), (-1, -4)\}$

6. $\{(0, 0), (-6, -2)\}$

7. $\{(3, -2), (-2, 3)\}$

8. $\{(3, 8), (-4, -6)\}$

9. $\left\{\left(\dfrac{5}{2}, -4\right), (2, -5)\right\}$

10. $\{(2, 5), (5, 2)\}$

Objective 2

11. $\{(1, 3), (1, -3), (-1, 3), (-1, -3)\}$

12. $\{(5, 2), (5, -2), (-5, 2), (-5, -2)\}$

13. $\{(2, 3), (2, -3), (-2, 3), (-2, -3)\}$

14. $\{(2, 1), (2, -1), (-2, 1), (-2, -1)\}$

15. $\{(3, 1), (3, -1), (-3, 1), (-3, -1)\}$

16. $\{(2, 3), (2, -3), (-2, 3), (-2, -3)\}$

17. $\{(1, 1), (1, -1), (-1, 1), (-1, -1)$

18. $\left\{\left(\dfrac{\sqrt{70}}{10}, \dfrac{3\sqrt{10}}{5}\right), \left(\dfrac{\sqrt{70}}{10}, \dfrac{-3\sqrt{10}}{5}\right), \left(\dfrac{-\sqrt{70}}{10}, \dfrac{3\sqrt{10}}{5}\right), \left(\dfrac{-\sqrt{70}}{10}, \dfrac{-3\sqrt{10}}{5}\right)\right\}$

19. $\{(2, 1), (2, -1), (-2, 1), (-2, -1)\}$

20. $\{(2, 0), (-2, 0)\}$

Objective 3

21. $\{(6, 1), (1, 6), (-6, -1), (-1, -6)\}$

22. $\{(2, 1), (-2, -1), (i, -2i), (-i, 2i)\}$

23. $\left\{(1, 1), (-1, -1), \left(\sqrt{3}, \dfrac{\sqrt{3}}{3}\right), \left(-\sqrt{3}, -\dfrac{\sqrt{3}}{3}\right)\right\}$

24. $\{(4, 3), (-4, -3), (3i, -4i), (-3i, 4i)\}$

25. $\{(2, -3), (-2, 3), (3, -2), (-3, 2)\}$

26. $\{(1, -4), (-1, 4), (4, -1), (-4, 1)\}$

27. $\left\{(3, -2), (-3, 2), \left(\dfrac{2\sqrt{6}}{3}, -\dfrac{3\sqrt{6}}{2}\right), \left(-\dfrac{2\sqrt{6}}{3}, \dfrac{3\sqrt{6}}{2}\right)\right\}$

28. $\left\{(2, 1), (-2, -1), \left(\sqrt{2}\ \sqrt{2}\right), \left(-\sqrt{2}, -\sqrt{2}\right)\right\}$

29. $\{(2, 2), (-2, -2), (2i, -2i), (-2i, 2i)\}$

30. $\left\{(3, -1), (-3, 1), \left(\dfrac{\sqrt{2}}{2}, -3\sqrt{2}\right), \left(-\dfrac{\sqrt{2}}{2}, 3\sqrt{2}\right)\right\}$

12.4 Mixed Exercises

31. $\left\{\left(\dfrac{3}{2}, \dfrac{1}{2}\right), \left(-\dfrac{3}{2}, -\dfrac{1}{2}\right)\right\}$

32. $\left\{\left(2, \dfrac{1}{2}\right), (1, 1)\right\}$

33. $\left\{\left(\sqrt{3}, 0\right), \left(-\sqrt{3}, 0\right)\right\}$

34. $\{(2, 1), (-2, -1), (i, -2i), (-i, 2i)\}$

35. $\{(3, 9), (-2, 4)\}$

36. $\{(2, 1), (2, -1), (-2, 1), (-2, -1)\}$

37. $\{(-2, 1), (2, -1), (i, 2i), (-i, -2i)\}$

38. $\left\{\left(\dfrac{3+\sqrt{5}}{2}, \sqrt{5}\right), \left(\dfrac{3-\sqrt{5}}{2}, -\sqrt{5}\right)\right\}$

39. $\left\{\left(\dfrac{i\sqrt{10}}{2}, -i\sqrt{10}\right), \left(\dfrac{-i\sqrt{10}}{2}, i\sqrt{10}\right), \left(\sqrt{5}, \sqrt{5}\right), \left(-\sqrt{5}, -\sqrt{5}\right)\right\}$

40. $\left\{\left(\sqrt{2}, -\sqrt{2}\right), \left(-\sqrt{2}, \sqrt{2}\right)\right\}$

12.5 Second-Degree Inequalities and Systems of Inequalities

Objective 1

1. $x \geq y^2$

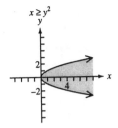

4. $25y^2 \leq 100 - 4x^2$

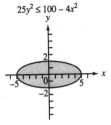

7. $x \leq 2y^2 + 8y + 9$

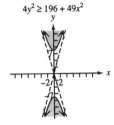

2. $y^2 \geq 9 - x^2$

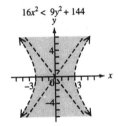

5. $x^2 + 4y^2 > 4$

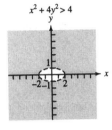

8. $4y^2 \geq 196 + 49x^2$

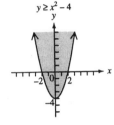

3. $16x^2 < 9y^2 + 144$

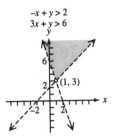

6. $y \geq x^2 - 4$

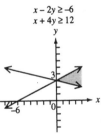

Objective 2

9. $-x + y > 2$
$3x + y > 6$

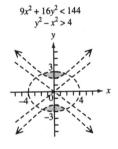

11. $x - 2y \geq -6$
$x + 4y \geq 12$

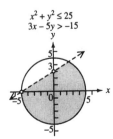

13. $9x^2 + 16y^2 < 144$
$y^2 - x^2 > 4$

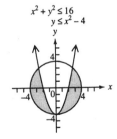

10. $x + y > -2$
$2x - y \leq -4$

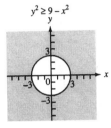

12. $x^2 + y^2 \leq 25$
$3x - 5y > -15$

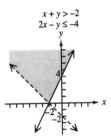

14. $x^2 + y^2 \leq 16$
$y \leq x^2 - 4$

12.5 Mixed Exercises

15. $7x^2 \le 42 - 6y^2$

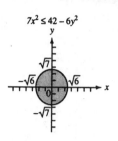

17. $x^2 > 9 - y^2$
$x \le 0$ and $y \ge 0$

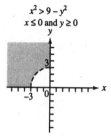

19. $4y + x^2 < 0$
$x \ge 0$

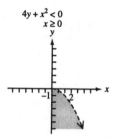

16. $9x^2 + 64y^2 \le 576$
$x \ge 0$

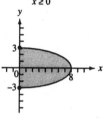

18. $x^2 - y^2 \le 16$
$y \ge 0$

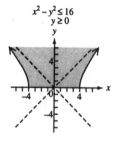

20. $x^2 + 4y^2 \le 36$
$-5 < x < 2$ and $y \ge 0$

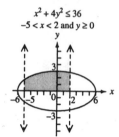